Machine Learning under Malware Attack

Raphael Labaca-Castro

Machine Learning under Malware Attack

Raphael Labaca-Castro
München, Germany

The following text is a reprint of a dissertation with homonymous title submitted to the Universität der Bundeswehr München in 2022.

ISBN 978-3-658-40441-3   ISBN 978-3-658-40442-0   (eBook)
https://doi.org/10.1007/978-3-658-40442-0

This Springer Vieweg imprint is published by the registered company Springer Fachmedien Wiesbaden GmbH, part of Springer Nature.
The registered company address is: Abraham-Lincoln-Str. 46, 65189 Wiesbaden, Germany

*For the loving souls who inspired it
but will never read it.*

Acknowledgments

I would like to sincerely thank everyone who supported me throughout this adventure. This PhD has been an amazing journey.

First and foremost, I am deeply grateful to my *Doktormutter*, Prof. Dr. Gabi Dreo Rodosek, for trusting me to join her team, for providing me with all the support and freedom I needed during my research, for connecting me with excellent professionals, and above all for always fostering a positive attitude when needed.

I would like to thank my co-advisor, Prof. Dr. Lorenzo Cavallaro, for his expertise, experience, and great advice for my research, as well as for bringing joy to our meetings and for the ice breakers and pleasant conversations during the cold times of video calls. I would also like to thank him for agreeing to work with me even after I said that all pasta is the same.

I would like to thank my colleagues and friends at the Research Institute CODE, especially Nils and Klement, with whom this PhD was very fun. I would like to thank them for our countless technical discussions, even though the three of us worked in different research areas. Over these years, I learned so much from other fields as I did from my own, and that was mostly because of you. I would also like to sincerely thank Volker and the university staff, who have been constantly helping me since the beginning of this journey to navigate this process with all resources needed and to bypass any administrative hurdles.

Thanks are due to Google X for hosting me twice as an AI resident in both California and Munich, to Joe for allowing me to gain his attention, and to everyone that I have worked with. I am really grateful for this opportunity to collaborate with world-class researchers. Working with you made me realize how intellectual humility seems to be related to acquired knowledge.

Thanks are also due to all colleagues and collaborators, to Christian and Jessica for being my first publication team, to Corinna for helping me achieve clarity throughout my first steps in PhD life, to Luis for the many interesting discussions around machine learning, and to Fabio and Feargus for the countless video calls we had while preparing our research. I hope we all meet personally soon in London or Munich. Thanks to Juanma and Fran who took the time to review the abstracts. I would also like to thank Marc and my team at work, who showed me nothing but patience and support while I was finalizing my dissertation and working full time.

Though not directly connected only to this PhD, I would like to thank my dear friends, especially those in Paysandú, Uruguay. Despite the distance, you have been there for me as a second family, and I am extraordinarily fortunate for having you in my life.

This work would not have been possible without my loving wife, Bel. Thank you so much for letting me share the adventure of life with you. Thank you for the frequent laughter, our road trips with *mates*, and for listening to me speak about my research on almost a daily basis. I could never imagine better companionship throughout these years.

Thank you Su for changing my life since I invaded yours, and thank you Lache, who is no longer with us but would have asked me to fix the television remote with my new degree. Thank you for the many lessons, especially for helping me understand that responsibility and discipline may lead to achieve great things, but at the end nothing will be more important than a united family.

Finally, I would like to particularly thank my parents for giving me life, love, grit, and strength, for raising me in multiple places, for kick-starting my passion toward discovering the world, for ultimately contributing to my *Fernweh*, and for supporting me in tough moments when I was unsure of myself but somehow you were. No effort will ever compare to what my family did for me to be here today.

I would like to dedicate a few last words to those who are now starting their academic journey. There are many reasons to not pursue a PhD. However, being able to explore new knowledge is an inspiring endeavor that is worth chasing. Learning to learn is perhaps the most underestimated task, yet it is highly useful to succeed in research and life.

While I have always imagined a PhD degree would be the end, I understand now that it is only the beginning, and I am glad to finally notice that there is no end.

Abstract

Machine Learning (ML) has become key in supporting decision-making processes across a wide array of applications, ranging from autonomous vehicles and streaming recommendations to network traffic analysis. Because of the massive amounts of information available, researchers and practitioners have largely focused on improving the performance of ML models. These efforts led to novel applications that can now overcome human performance nearly in every domain.

Despite being highly accurate, these algorithms have been shown to be prone to returning incorrect predictions. This means that the trustworthiness of a model may be compromised in terms of both training and testing time. In other words, committed adversaries may be able to manipulate input data (e.g., images and network traffic) to resemble objects from another class, performing what is known as *evasion* attacks. They may also use another strategy called *poisoning*, which includes injecting undesirable data into the training set. Both approaches aim to mislead the model to predict the wrong label for a given object.

While adversarial attacks may target how a model is induced to predict a certain class for an object (e.g., classifying a red traffic light as green), this is normally sufficient with a wrongly predicted label or the opposite class for binary classification, which is generally the case in the context of malware (e.g., classifying malicious software as harmless).

In the event of manipulating objects for evasion attacks, these are known as *adversarial examples* and pose multiple security risks. On many occasions, as in malware classification, such behavior needs to be further examined to assess the degree to which predictions can be trusted.

Therefore, studying adversarial attacks can help identify systemic weaknesses in ML classifiers. These attacks reveal weak spots in the model that allow carefully manipulated objects to be incorrectly classified, hence compromising the quality of predictions. In fact, by investigating multiple strategies to generate successful adversarial examples, models can be evaluated from multiple perspectives and potentially hardened against adversarial examples. However, it is worth noting that while attacks generally materialize in the feature domain and are convertible to the problem space, where they exist in the real world, this is not always the case in the context of malware, especially in Portable Executable (PE) files. In this context, generating real adversarial malware examples often requires modifications that preserve binary integrity at the byte level. Thus, creating effective attacks using PEs is not a trivial task.

In this study, we present a framework that contains a suite of input-specific attacks using PE malware targeting ML-based classifiers, in which the adversaries have limited knowledge about target classifiers. We explore multiple approaches in which the adversary leverages hard labels from the model and does not have any prior knowledge of the architecture or access to the training data. To deeply understand the model's behavior, we additionally study full-knowledge attacks based on gradient information.

We introduce universal adversarial attacks in the problem space for PEs. The underlying goal here is to show whether the generation of adversarial examples can be automated and generalized without relying exclusively on input-specific attacks to generate effective adversarial examples.

We also propose a defense strategy that leverages knowledge from the aforementioned attacks to increase the cost of generating adversarial examples and, therefore, improve the target model against carefully crafted objects produced by adaptive adversaries. We envision a holistic approach that facilitates the identification of systemic vulnerabilities and enhances the classifier's resilience at a reasonable cost.

Next, we perform a statistical analysis of malware features by evaluating the impact of real-world attacks in the feature domain, which provides clarity for model predictions under unexpected input.

Finally, we release our Framework for Adversarial Malware Evaluation and make the source code available to encourage participation and further research into this fascinating topic and promote the evaluation and building of more resilient malware classifiers.

Kurzfassung

Maschinelles Lernen (ML) ist zu einem zentralen Baustein für die Unterstützung von Entscheidungsprozessen in einer Vielzahl von Anwendungen geworden, die von autonomen Fahrzeugen über Streaming-Empfehlungen bis hin zur Analyse des Netzverkehr reichen. Aufgrund der großen Menge an verfügbaren Informationen haben sich Forscher und Praktiker weitgehend auf die Verbesserung der Leistung von ML-Modellen konzentriert. Diese Bemühungen haben zu neuartigen Anwendungen geführt, die nun in fast jedem Bereich die menschliche Leistung übertreffen können.

Es hat sich gezeigt, dass diese Algorithmen trotz ihrer hohen Genauigkeit dazu neigen, falsche Vorhersagen zu machen. Dies bedeutet, dass die Vertrauenswürdigkeit eines Modells sowohl in Bezug auf die Trainings- als auch die Testzeit beeinträchtigt werden kann. Mit anderen Worten: Engagierte Angreifer können die Eingabedaten (z. B. Bilder und Netzverkehr) so manipulieren, dass sie Objekten einer anderen Klasse ähneln, und damit sogenannte *evasion*-Angriffe durchführen. Sie können auch eine andere Strategie anwenden, welche als *poisoning* bezeichnet wird und das Einspeisen unerwünschter Daten in den Trainingsdatensatz beinhaltet. Beide Ansätze zielen darauf ab, das Modell dazu zu verleiten, die falsche Bezeichnung für ein bestimmtes Objekt vorherzusagen.

Feindliche Angriffe können darauf abzielen, wie ein Modell dazu gebracht wird, eine bestimmte Klasse für ein Objekt vorherzusagen (z. B. eine rote Ampel als grün zu klassifizieren). Es reicht hierbei normalerweise aus, wenn ein falsches Label oder, bei einer binären Klassifizierung, die entgegengesetzte Klasse vorhergesagt wird, was im Allgemeinen im Zusammenhang mit Malware der Fall ist (z. B. die Klassifizierung von bösartiger Software als harmlos).

Manipulierte Objekte für Ausweichangriffe werden als *adversarial examples* bezeichnet und stellen mehrere Sicherheitsrisiken dar. In vielen Fällen, wie z. B.

bei der Klassifizierung von Schadsoftware, müssen diese manipulierten Objekte weiter untersucht werden, um zu beurteilen, inwieweit den Vorhersagen vertraut werden kann.

Die Untersuchung feindlicher Angriffe kann daher dazu beitragen, systemische Schwachstellen in ML-Klassifikatoren zu ermitteln. Diese Angriffe decken Schwachstellen im Modell auf, die es ermöglichen, dass sorgfältig manipulierte Objekte falsch klassifiziert werden, wodurch die Qualität der Vorhersagen beeinträchtigt wird. Durch die Untersuchung mehrerer Strategien zur Generierung erfolgreicher adversarial examples können Modelle aus verschiedenen Blickwinkeln bewertet und potenziell gegen adversarial examples abgehärtet werden. Es ist jedoch anzumerken, dass Angriffe in der Regel in der Merkmalsdomäne (*feature-space*) durchgeführt und anschließend in den Problemraum (*problem-space*) konvertiert werden, in dem sie in der realen Welt vorkommen. Dies ist im Zusammenhang mit Malware, insbesondere bei Portable Executable (PE)-Dateien, nicht immer der Fall. Die Generierung von echten adversarial malware examples erfordert oft Änderungen, welche die binäre Integrität auf Byte-Ebene bewahren. Daher ist die Erstellung effektiver Angriffe mit PEs keine triviale Aufgabe.

In dieser Studie stellen wir ein Framework vor, welches eine Reihe von eingabespezifischen Angriffen mit PE-Malware enthält. Diese zielen auf ML-basierte Klassifikatoren ab, wobei die Angreifer nur begrenzte Kenntnisse über die Zielklassifikatoren haben. Wir untersuchen mehrere Ansätze, bei denen der Angreifer feste Kennzeichnungen des Modells ausnutzt und keine Vorkenntnisse über die Architektur oder Zugriff auf die Trainingsdaten hat. Um das Verhalten des Modells besser zu verstehen, untersuchen wir zusätzlich Angriffe bei vollständigem Wissen, die auf Gradienteninformationen basieren.

Wir stellen universelle gegnerische Angriffe im Problemraum für PEs vor. Das zugrunde liegende Ziel ist es, zu zeigen, ob die Generierung von adversarial examples automatisiert und verallgemeinert werden kann, ohne sich ausschließlich auf eingabespezifische Angriffe zu verlassen, um effektive adversarial examples zu generieren.

Wir schlagen außerdem eine Verteidigungsstrategie vor, die das Wissen aus den oben erwähnten Angriffen nutzt, um die Kosten für die Generierung von feindlichen Angriffen zu erhöhen und somit das Zielmodell gegen sorgfältig ausgearbeitete Objekte zu verbessern, die von adaptiven Gegnern erzeugt werden. Wir stellen uns einen ganzheitlichen Ansatz vor, der die Identifizierung von systemischen Schwachstellen erleichtert und die Widerstandsfähigkeit des Klassifikators zu vertretbaren Kosten erhöht.

Als Nächstes führen wir eine statistische Analyse von Malware-Merkmalen durch, indem wir die Auswirkungen realer Angriffe im Merkmalsdomäne bewerten, was Klarheit für Modellvorhersagen bei unerwarteten Eingaben schafft.

Abschließend veröffentlichen wir unser Framework for Adversarial Malware Evaluation und stellen den Quellcode zur Verfügung, um weitere Forschung zu diesem faszinierenden Thema zu fördern und die Evaluierung und Entwicklung von widerstandsfähigeren Malware-Klassifikatoren zu unterstützen.

Resumen

El aprendizaje automático, en inglés conocido como machine learning (ML), se ha convertido en un elemento clave para apoyar los procesos de toma de decisiones en un amplio abanico de aplicaciones, que van desde los vehículos autónomos y las recomendaciones en streaming hasta el análisis del tráfico de la red. Debido a la enorme cantidad de información disponible, los investigadores y profesionales se han centrado en gran medida en mejorar el rendimiento de los modelos de ML. Estos esfuerzos han dado lugar a nuevas aplicaciones que ahora pueden superar el rendimiento humano en casi todos los ámbitos.

A pesar de ser muy precisos, estos algoritmos han demostrado ser propensos a realizar predicciones incorrectas. Esto significa que la fiabilidad de un modelo puede verse comprometida tanto en el tiempo de entrenamiento como en el de prueba. En otras palabras, los adversarios pueden ser capaces de manipular los datos de entrada (por ejemplo, imágenes y tráfico de red) para que se parezcan a objetos de otra clase, realizando lo que se conoce como ataques de *evasión*. También pueden utilizar otra estrategia denominada *envenenamiento*, que incluye la inyección de datos no deseados en el conjunto de entrenamiento. Ambos enfoques tienen como objetivo inducir al modelo a predecir una clasificación incorrecta para un objeto determinado.

Para que los ataques adversarios logren el objetivo de inducir a un modelo a predecir una determinada clase para un objeto (por ejemplo, clasificar un semáforo rojo como verde), normalmente es suficiente con una etiqueta predicha erróneamente o la clase opuesta para la clasificación binaria, que es generalmente el caso en el contexto del malware (por ejemplo, clasificar código malicioso como inofensivo).

En el caso de la manipulación de objetos para ataques de evasión, éstos se conocen como *ejemplos adversarios* y suponen múltiples riesgos de seguridad. En muchas ocasiones, como en la clasificación de malware, es necesario examinar más a fondo este tipo de comportamiento para evaluar el grado en que se puede confiar en las predicciones.

Por lo tanto, el estudio de los ataques adversarios puede ayudar a identificar las debilidades sistémicas de los clasificadores de ML. Estos ataques revelan puntos débiles en el modelo que permiten clasificar incorrectamente objetos cuidadosamente manipulados, comprometiendo así la calidad de las predicciones. De hecho, al investigar múltiples estrategias para generar ejemplos adversos exitosos, los modelos pueden ser evaluados desde múltiples perspectivas y potencialmente fortalecidos contra ejemplos adversarios. Sin embargo, cabe señalar que, si bien los ataques suelen materializarse en el dominio de las características (*feature space*) y son convertibles al dominio del problema (*problem space*), donde existen en el mundo real, no siempre es así en el contexto del malware, especialmente en los archivos ejecutables portátiles (EP). En este contexto, la generación de ejemplos reales de malware adverso suele requerir modificaciones que preserven la integridad binaria a nivel de bytes. Por lo tanto, crear ataques eficaces utilizando EP no es una tarea trivial.

En este estudio, presentamos un *framework* que contiene un conjunto de ataques específicos de entrada (*input-specific*) utilizando malware EP dirigido a clasificadores basados en ML, sobre los cuales los adversarios tienen un conocimiento limitado. Exploramos múltiples enfoques en los que el adversario aprovecha las etiquetas duras (*hard labels*) del modelo sin tener ningún conocimiento previo de la arquitectura ni acceso a los datos de entrenamiento. Para comprender en profundidad el comportamiento del modelo, estudiamos además los ataques de conocimiento completo basados en la información de gradiente.

Formulamos ataques adversarios universales en el espacio de problemas de los EP. El objetivo subyacente aquí es mostrar si la generación de ejemplos adversarios puede ser automatizada y generalizada sin depender exclusivamente de ataques específicos de entrada para generar ejemplos adversarios efectivos.

También proponemos una estrategia de defensa que aprovecha el conocimiento de los ataques antes mencionados para aumentar el coste de la generación de ejemplos adversarios y, por lo tanto, mejorar la seguridad del modelo objetivo contra los objetos cuidadosamente elaborados producidos por adversarios adaptativos. La intención es generar un enfoque holístico que facilite la identificación de vulnerabilidades sistémicas y mejore la resistencia del clasificador a un coste razonable.

Posteriormente, realizamos un análisis estadístico de las características del malware evaluando el impacto de los ataques del mundo real en el dominio de las características, lo que proporciona claridad para las predicciones del modelo bajo entradas inesperadas.

Por último, publicamos nuestro Framework for Adversarial Malware Evaluation y ponemos a disposición el código fuente para fomentar la participación y la investigación de este fascinante tema y promover la evaluación y construcción de clasificadores de malware más resistentes.

Resumo

O aprendizado de máquina, em inglês machine learning (ML), tornou-se fundamental no apoio aos processos de tomada de decisão numa vasta gama de aplicações, desde veículos autônomos e recomendações de streaming até à análise do tráfego na rede. Devido à enorme quantidade de informação disponível, os pesquisadores e profissionais concentraram-se em grande parte na melhoria do desempenho dos modelos ML. Estes esforços conduziram a novas aplicações que podem agora superar o desempenho humano em quase todos os domínios.

Apesar de serem altamente precisos, estes algoritmos têm demonstrado ser propensos a retornar previsões incorretas. Isto significa que a fiabilidade de um modelo pode ser comprometida tanto em termos de formação como de teste. Em outras palavras, os adversários comprometidos podem ser capazes de manipular dados de entrada (por exemplo, imagens e tráfego de rede) para se assemelharem a objetos de outra classe, realizando o que é conhecido como ataques de *evasão*. Eles podem também utilizar outra estratégia chamada *poisoning*, que inclui a injeção de dados indesejáveis no conjunto de dados treino. Ambas as abordagens visam enganar o modelo para prever a etiqueta errada para um determinado objeto.

Embora os ataques contraditórios possam visar a forma como um modelo é induzido a prever uma determinada classe para um objeto (por exemplo, classificar um semáforo vermelho como verde), isto é normalmente suficiente com uma etiqueta mal prevista ou a classe oposta para a classificação binária, o que é geralmente o caso no contexto de malware (por exemplo, classificar um software malicioso como inofensivo).

No caso de manipulação de objetos para ataques de evasão, estes são conhecidos como *exemplos adversários* e representam múltiplos riscos de segurança.

Em muitas ocasiões, como na classificação de malware, tal comportamento precisa ser examinado mais detalhadamente para avaliar o grau em que as previsões podem ser confiáveis.

Portanto, o estudo de ataques adversos pode ajudar a identificar fraquezas sistémicas nos classificadores ML. Estes ataques revelam pontos fracos no modelo que permitem a classificação incorreta de objetos cuidadosamente manipulados, comprometendo assim a qualidade das previsões. De fato, ao investigar múltiplas estratégias para gerar exemplos adversários bem-sucedidos, os modelos podem ser avaliados de múltiplas perspectivas e potencialmente fortalecidos contra exemplos adversários. No entanto, vale a pena notar que embora os ataques se materializem geralmente no domínio das características (*feature space*) e sejam convertíveis no espaço do problema (*problem space*), onde existem no mundo real, nem sempre é esse o caso no contexto do malware, especialmente nos arquivos Portable Executable (PE). Neste contexto, gerar exemplos reais de malware adversário requer frequentemente modificações que preservem a integridade binária ao nível do byte. Assim, a criação de ataques eficazes utilizando PE não é uma tarefa trivial.

Neste estudo, apresentamos um quadro que contém um conjunto de ataques específicos de entrada usando malware de PE que visam classificadores baseados em ML, no qual os adversários têm um conhecimento limitado sobre classificadores-alvo. Exploramos múltiplas abordagens nas quais o adversário aproveita etiquetas duras (*hard labels*) do modelo e não tem qualquer conhecimento prévio da arquitectura ou acesso aos dados de formação. Para compreender profundamente o comportamento do modelo, estudamos adicionalmente os ataques de conhecimento completo com base em informação de gradiente.

Formulamos ataques adversários universais no espaço do problema para os PEs. O objectivo aqui é mostrar se a geração de exemplos de adversários pode ser automatizada e generalizada sem depender exclusivamente de ataques específicos de entrada para gerar exemplos efectivos de adversários.

Propomos também uma estratégia de defesa que aproveita o conhecimento dos ataques mencionados anteriormente para aumentar o custo da geração de exemplos adversários e, portanto, melhorar o modelo alvo contra objetos cuidadosamente trabalhados produzidos por adversários adaptativos. A intenção é realizar uma abordagem holística que facilite a identificação de vulnerabilidades sistémicas e aumente a resiliência do classificador a um custo razoável.

Posteriormente, realizamos uma análise estatística das características do malware, avaliando o impacto dos ataques do mundo real no domínio das características, o que proporciona clareza para as previsões do modelo sob entrada inesperada.

Finalmente, divulgamos o nosso Framework for Adversarial Malware Evaluation e disponibilizamos o código fonte para fomentar a participação e investigação adicional sobre este fascinante assunto e promover a avaliação e construção de classificadores de malware mais resistentes.

Contents

Part VI   Closing Remarks

Acronyms

Framework-specific

AIMED	Automatic Intelligent Manipulations to Evade Detection
AIMED-RL	Automatic Intelligent Manipulations to Evade Detection with RL
ARMED	Automatic Random Manipulations to Evade Detection
FAME	Framework for Adversarial Malware Evaluation
GAME-UP	Generate Adversarial Malware Examples with Universal Perturbations
GAINED	Generative Adversarial Intelligent Network to Evade Detection
GRIPE	Gradient Relevant Injection to Portable Executables

Framework-agnostic

ACER	Actor Critic with Experience Replay
AML	Adversarial Machine Learning
APK	Android Package Kit
CFG	Control Flow Graph
CNN	Convolutional Neural Network
COFF	Common Object File Format
DDQN	Double DQN
DiDQN	Distributional DQN
DiDDQN	Distributional Double DQN
DLL	Dynamic Link Library
DQN	Deep Q-Network
FGSM	Fast Gradient Sign Method

FNR	False Negative Rate
FPR	False Positive Rate
GAN	Generative Adversarial Network
GBDT	Gradient Boosted Decision Tree
GNN	Graph Neural Network
GP	Genetic Programming
IDS	Intrusion Detection System
IAT	Import Address Table
IPS	Intrusion Prevention System
LightGBM	Light Gradient Boosting Machine
LK	Limited Knowledge
LR	Logistic Regression
MDP	Markov Decision Process
ML	Machine Learning
MTD	Moving Target Defense
OS	Operating System
PE	Portable Executable
PDF	Portable Document Format
PK	Perfect Knowledge
RL	Reinforcement Learning
SVM	Support Vector Machine
TPR	True Positive Rate
UAP	Universal Adversarial Perturbation
UER	Universal Evasion Rate
UPX	Ultimate Packer for eXecutables

List of Figures

List of Tables

Part I
The Beginnings of Adversarial ML

Introduction 1

All truths are easy to understand once they are discovered; the point is to discover them.
—Galileo Galilei

Generally, it is difficult to imagine the world today without the influence of Machine Learning (ML) from improving medical diagnosis [RLJ20] to autonomous vehicles [Jan+20]. Although ML models have been ubiquitously deployed to make life easier, not all of the algorithms have been vetted enough to ensure their safety, which is an often neglected aspect when designing solutions.

Instead, performance is considered to be paramount not only from a market perspective, but also in the research arena. Articles submitted to ML conferences are deeply interested in pushing the state of the art through expensive training stages for very large models [Bro+20; Ng+21; Zha+20], often in exchange for a marginal improvement. Likewise, many practitioners are fascinated by rapid scaling, which causes important aspects of models to be left out of the equation, such as in the cases of algorithmic discrimination, energy consumption, public safety, and security concerns [GS18; Lit+21].

1.1 Motivation

These problems in ML models, especially the security issues, are domain-agnostic, with varying degrees of susceptibility across different fields. While autonomous vehicles may misclassify a stop traffic sign and unexpectedly drive into the traffic [Qay+20], network security models can flag malicious behavior as benign, compromising both users and organizations [Ibi+19]. Therefore, evaluating the

© The Author(s), under exclusive license to Springer Fachmedien Wiesbaden GmbH, part of Springer Nature 2023
R. Labaca-Castro, *Machine Learning under Malware Attack*,
https://doi.org/10.1007/978-3-658-40442-0_1

weaknesses of ML models is important before they are deployed into production, since calculated attacks can pose a critical risk to security (e.g., data) and safety, which refers to the feeling of being free from any danger caused by ML algorithms.

Given the tremendous impact of such models on daily life, different organizations may not be making enough effort toward analyzing such concerns when training ML algorithms. Indeed, identifying novel approaches in the research field represents a challenge, as the number of publications has been exponentially increasing, with more than 2,000 papers published since 2014 [Kum+20]. Furthermore, researchers often limit access to their work or propose novel solutions without reproducible experiments and detailed implementations.

For all of these reasons, securing ML models in the sense of guaranteeing trust in their predictions remains an open challenge in the field, despite the numerous approaches to improve robustness [Car+19]. The issue here lies in the fact that covering vulnerability pockets from the data distribution may create new weaknesses. For example, adjusting a model to detect adversarially modified objects may cause it to overlook genuine input, which compromises its overall performance. Hence, models should be properly evaluated by confronting them with *adversarial examples*, modified input objects that are created to induce errors in the classifier's predictions. However, there is a risk of evaluating models against weak attacks [Ues+18] and even legal implications of Adversarial Machine Learning (AML) that may need to be addressed [Kum+20].

Overall, these shortcomings are considered an opportunity for adversaries to profit from low levels of collaboration between researchers and practitioners, the lack of benchmarks in the domain, and the complexity of creating robust defenses.

1.2   Problem Statement

While intense work toward improving performance metrics has led to highly accurate models, ML algorithms may still exhibit unexpected behavior that may cause them to associate input objects with wrong labels [Bar+06; Big+13; Hua+11; KL93; Sze+13]. In other words, targeted models can make a wrong decision when confronted with manipulated objects. Such behavior is consistent within a variety of settings and contexts [BR18; Cha+18; GBC16].

Among all areas, security is considered to be particularly interesting, since adversaries explore the means of maximizing profit using different strategies, including malware propagation and opportunities in the underground market [And+19; CW19; Fos+08]. While the cybersecurity field is *per se* dedicated to protection against such adversaries, the adversarial attacks themselves consist of a different conundrum. The same powerful technology that aims to improve models' predictions is the one putting them at risk of novel attacks. This issue has clear implications, especially in the context of malware, in which malicious software can be used to perpetrate attacks targeting not only traditional users [FN17], but also more unconventional targets, such as uranium enrichment facilities [BR17] or genomic data [Lab15]. In this context, an adversary may design a carefully crafted set of transformations in the form of a toolbox to inject them into previously detected objects. If successful, malicious actors can then force ML models to misclassify the adversarial object as benign without flagging the risk [Kol+18; Pie+20; SCJ19].

However, there is another problem in the context of malware that remains to be solved. That is, the binary files can be corrupted if the transformations[1] drastically change their structure to persuade the models to flip the label. While adversarial attacks lack such restrictions in particular domains, such as in image processing, software binaries, specifically Windows Portable Executable (PE) files, may need to be further reviewed using a verification stage to ensure that the *plausibility* of the object is preserved [Pie+20].

This work seeks to address the aforementioned shortcomings of AML in the context of malware with PE files. However, to better understand how models behave with adversarial input, it is important to define a systemic approach that allows evaluating adversarial attacks against malware classifiers through an automated pipeline. Another important aspect is to produce valid adversarial examples efficiently in the context of malware. As previously discussed, attacks are required to modify input objects to produce effective adversarial malware while retaining the object's integrity. Hence, identifying efficient strategies to produce realizable adversarial objects is essential. Furthermore, effective attacks show vulnerabilities in the target models, which may lead to the generation of enhanced algorithms. However, improving the resilience of classifiers without compromising the detection of genuine objects is a challenging task that relies on the nature of PE malware and differs from related domains. Based on these limitations, we hereby summarize the research questions that we plan to answer:

---

[1] Note that, in the literature, the terms *perturbation* and *transformation* generally refer to noise added to the input object and may differ when it occurs in the *feature-space* or the *problem-space*. While we are aware of the differences, since our research mostly focuses on real byte-level perturbations, we will employ both terms indistinguishably.

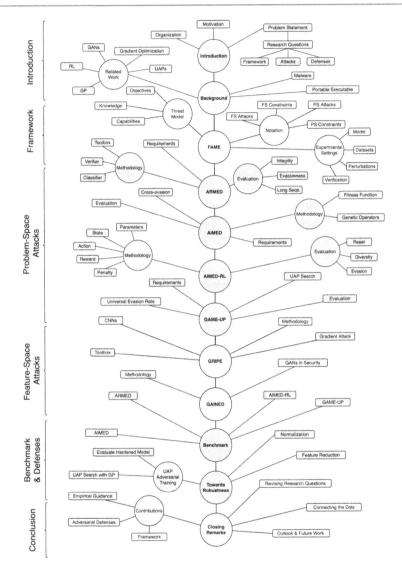

Figure 1.1 Structure of the Dissertation. Modules are grouped using white or gray colors and represented at different levels of hierarchy. For example, as depicted in white, *Objectives*, *Knowledge*, and *Capabilities* are components (subsections) of the *Threat Model* (section) included in the *FAME* (chapter) inside the *Framework* (part)

RQ1: How can we define a systematic approach that allows weaknesses of malware models to be methodologically assessed using real-world attacks? (Part II)

RQ2: To what extent can the generation of adversarial examples against malware classifiers be efficiently automated while remaining plausible? (Part III & IV)

RQ3: Are there effective universal attacks in the context of malware and how can these enhance a model's resilience against adaptive adversaries? (Part V)

1.3 Organization of Dissertation

The remainder of this work is structured as follows. In Chapter 2, we define the background and present preliminary knowledge, including a brief introduction to malware and the specifications of the PE format. Then, we investigate the literature on AML in the context of malware. In Chapter 3, we introduce our framework and define the scope of the problem, including the adversaries' goals and knowledge, and present the target models along with malware datasets. In Chapters 4–7, we introduce multiple attack strategies in the problem space, whose goal is to leverage stochastic and ML techniques to efficiently identify weaknesses in the target models, in which adversaries have restricted knowledge, resembling scenarios closer to the real world. Likewise, in Chapters 8 and 9, we propose full-knowledge attacks in the feature space using gradient optimization. With this white-box scenario, the goal is to efficiently bypass the target model with malware representations that respect real binary transformations. In Chapter 10, we present a comparison among the proposed realizable attacks using a separate evaluation scenario and then benchmark their performance by analyzing the best use cases for each strategy. Next, in Chapter 11, we present multiple strategies to improve the resilience of static malware classifiers on the basis of knowledge leveraged from the attack modules. Finally, in Chapter 12, we review the contributions of this study as a conclusion and provide an outlook for future work (Fig. 1.1).

Background

2

Most of the fundamental ideas of science are essentially simple, and may, as a rule, be expressed in a language comprehensible to everyone.
—Albert Einstein

In this section, we address some important concepts that form the basis for our research and are required for the understanding of this work. We start by defining the origin of malicious applications to understand their impact. Next, we present the PE format, which is the type of binary that we use as an input object during our experimental evaluation. We then explore AML and review the literature, starting with early implementations in the security domain, ranging from spam filtering to the malware classification problem.

2.1 Malware

Computer viruses were formally introduced by Fred Cohen [Coh86; Coh87] in the second half of the 1980s. In his studies, Cohen explored the computational aspects of computer viruses and their mathematical properties. Initially, computer viruses were defined as programs that impact and modify similar programs by including a version of itself. This definition, however, does not necessarily embody a malicious attempt. Therefore, we decided to adopt Radai's approach [Rad89], since we assume that adversaries create such programs to compromise systems as a goal, even if this occurs accidentally during experiments [SH82]. Besides attacks, Cohen investigated practical defenses against computer viruses [Coh89], and Kephart et al. [Kep+95] introduced biologically inspired early defensive measures.

© The Author(s), under exclusive license to Springer Fachmedien Wiesbaden GmbH, part of Springer Nature 2023
R. Labaca-Castro, *Machine Learning under Malware Attack*,
https://doi.org/10.1007/978-3-658-40442-0_2

Malware [Rad89] is a portmanteau of *mal*icious soft*ware*, that refers to any type of malicious code, including computer viruses, worms, and Trojan horses [AP95]. Many attackers regard such threats as a promising tool to generate income from spam used to advertise illicit products to malicious software that propagates phishing pages with the intention of stealing login credentials [Fos+08]. As a result, adversaries are constantly motivated to further develop their threats, in some cases using the same technology deployed to detect them, such as ML [PY20].

While malware can be created for virtually any Operating System (OS), in this work, we will largely focus on PE software, since Microsoft Windows is the most used platform [Dep21]. Thus, it makes more sense for adversaries to direct their effort toward the most prominent targets because of their popularity [AV-20].

2.2   Portable Executable Format

The PE format was introduced to define a common system to run applications on the entire Windows family in order to store the information required to execute software across supported architectures [Pie94]. In fact, the word *portable* refers to the agnostic nature of the file toward the architecture. Besides belonging to the original Win32 specifications, PE files extend the Common Object File Format (COFF) and include Dynamic Link Library (DLL) files as well. The only difference between both of them is a bit that denotes whether the file is EXE or DLL.

As can be observed in Fig. 2.1, the PE format starts with *header* information, in which the first fields *DOS header* and *DOS stub* are legacy portions used for back compatibility. The fields within the *PE header* provide information to the OS about how to map the file into memory, since not all the PE is mapped. For example, relocations may be read but not mapped, whereas parts at the end of the file may not even be mapped. The *PE optional header* provides additional information about the file, such as its size and entry point address. At the end of the header, there is a *section table*, which is represented by an array structure that includes information about each section defined in the PE. This means that the number of entries in the table should match the number of sections in the file.

Following the header information are the *sections* of the file that are responsible for grouping the code and data required for execution. The first section is the *.text* section, in which the executable code is allocated into one large section along with pointers to the import table. The *.data* section contains read-only information (e.g., constants and literal strings) that do not change during execution. Along these sections, there are uninitialized variables that unnecessarily increase the size of the file if declared and allocated in the executable. The *.idata* section stores information about imported functions, including the Import Address Table (IAT) and import directory, whereas the *.edata* section allocates the export directory, which includes exported functions' names and addresses. Although the last section depicted in Fig. 2.1 is the *.rsrc* section, which includes resource information (e.g., icons and strings), the PE file contains further sections and unmapped data, such as debug information and *overlay*, which normally represent the last part of the PE file.

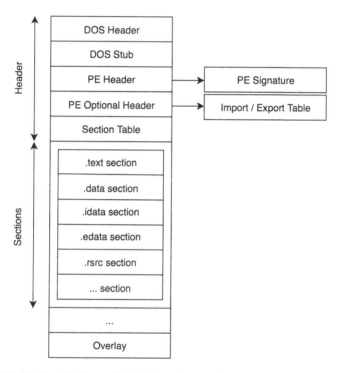

Figure 2.1 The Portable Executable file format layout. The representation divides the PE in two parts: Header, that includes DOS and PE Header, which also contains the PE Signature; and the Sections, that stores the data mainly responsible for execution

Some attempts have been made to manipulate PE files [LS95; Sri93]. However, creating effective multipurpose approaches is a complex task, which is why researchers may opt to implement their own parsers that are connected to a specific language and context, making it harder to be reused. To address this shortcoming, a cross-platform library was created, called LIEF [Tho18] that aims to parse and provide an interface for manipulating executable files, including PEs.

2.3   Related Work

For almost two decades, ML models have been investigated and found to exhibit curious behavior [KL93]. For instance, Dalvi et al. [Dal+04] explored attacks that aim to minimize the distance to decision boundaries for linear classifiers. Later on, Lowd and Meek [LM05] investigated adversarial learning and proposed an approach for building adversarial examples based on information learned about the model.

Barreno et al. [Bar+06] explored an early taxonomy of adversarial attacks and their applications in email filtering and Intrusion Detection Systems (IDSs) that are designed to identify threats at a network level. Further approaches were proposed to extend the work addressing spam filtering [BFR10; GR06; KT+09; Nel+08], which was an important security concern in the early 2000s. Later, Barreno [Bar+10] also presented a comprehensive taxonomy analyzing attacks against ML models.

Šrndic and Laskov [ŠL13] shifted evasion attacks toward Portable Document Format (PDF) malware models and formulated the idea that nonlinear classifiers should be more secure. Biggio et al. [Big+13] showed that neural networks present weaknesses to gradient attacks, defining Perfect Knowledge (PK) and Limited Knowledge (LK) attacks from the adversary perspective. Concurrently, Szegedy et al. [Sze+13] introduced minimum-distance gradient attacks against neural networks processing images using the MNIST [LeC+98] dataset and also presented an early approach for adversarial training.

Huang et al. [Hua+11] introduced a formalization of AML and extended the taxonomy, including a categorization based on influence, security violation, and specificity. They presented two case studies using spam filtering and a network anomaly detector to show how vulnerable the models are to evasion attacks, which paved the way for further adversarial approaches targeting IDS devices [AC18; Apr+20; CGR13].

In the malware domain, adversarial examples have been found in a variety of platforms, including Windows [Kol+18; Ros+18], Android [Dem+17, Gro+16, Yan+17], PDF documents [Big+13; ŠL13; XQE16], and Javascripts [FBS19].

In this work, we seek to generate real binaries in the problem space and embed representations in the feature space of PE files on Windows. Since there are a number of strategies that include adversarial examples in the literature, we categorize the most prominent approaches under the following groups: gradient optimization, generative networks, reinforcement learning, genetic programming, and universal adversarial perturbations. Table 2.1 shows a comparison of the approaches.

2.3.1 Gradient Optimization

Biggio et al., [Big+13] performed an early study on malware evasion targeting PDF classifiers using gradient descent attacks, in which different levels of adversary knowledge were proposed, full or perfect (PK) and partial or limited (LK). In the case of PK, the attackers have access to the trained target model, type, and feature space, whereas in the case of LK, they know the type of classifier and feature space but have no access to the pretrained model or its training set. In this context, the goal of the adversary is to manipulate a single input object until it crosses the decision boundary. Therefore, an improved attack would be to maximize the confidence of the classifier when the adversarial examples are classified by minimizing the value of the model's discriminant function. Hence, the authors targeted a neural network and a Support Vector Machine (SVM) reporting high evasion probabilities, including perfect evasion for certain scenarios depending on the model, adversary knowledge, and the use of *mimicry* attacks, which they referred to as mimicking the characteristics of benign applications.

Kreuk et al. [Kre+18] introduced an approach using the Fast Gradient Sign Method (FGSM) [GSS15] to generate adversarial examples targeting MalConv [Raf+17], which is a Convolutional Neural Network (CNN) trained on raw binary sequences without domain-based feature engineering. According to the authors, the functionality of the files can be preserved by limiting the manipulations to a small set of bytes that are either placed in sections of unused bytes or appended at the end of the file as a new section. Then, behavior graphs are computed to verify that the files' actions remain the same after the payload is injected. As a result, the attacks revealed up to 99.21 % evasion rates for the best-case scenarios.

Al-Dujaili et al. [Al-+18] presented an attack inspired by linear programming relaxation that extends the malware binary domain to deterministic and randomized rounding, leading to FGSM derivations, that is, dFGSM and rFGSM, respectively. They implemented, as a malware classifier, a feed-forward neural network that consists of three hidden layers with 300 neurons each and two output nodes using the LogSoftMax function. Then, they trained the model with their own binary dataset, SLEIPNIR, which represents PE files in the feature space.

Kolosnjaji et al. [Kol+18] proposed a gradient attack that limits manipulations to less than 1 % of the adversarial example, which is sufficient to evade deep neural networks trained with raw bytes, such as MalConv. The approach included only one type of transformation, byte append, and relied on it exclusively to generate adversarial examples. As a result, the authors reported up to 60 % evasion rates.

Overall, gradient-based attacks leverage mostly PK scenarios in which the adversary has full access to the model and its hyperparameters and features. This results in predominantly high evasion rates. Another reason for the high performance observed is that preserving the integrity of objects in the feature space is significantly less complex than in the real world. Since no output binary object is produced, integrity verification is commonly achieved by restricting the value of the features modified, which is a less invasive technique compared to byte-level perturbations in the problem space.

2.3.2 Generative Adversarial Networks

Hu et al. [Goo+14] used Generative Adversarial Networks (GANs) to create Mal-GAN [HT17]. The goal of GANs is to deploy two neural networks that collaborate with each other: one known as the *generator*, which is responsible for generating adversarial examples, and the other known as the *discriminator* that learns how to detect the adversarial examples created by the first network. In their implementation, Hu et al. made the discriminator act as a surrogate model that fits a black-box classifier, since GANs require full visibility of the model (i.e., gradient information) to implement the attack. However, although the authors reported perfect evasion, the adversarial examples were produced only in the feature space, with no real objects generated in the process.

Rigaki et al. [RG18] proposed the use of GANs to learn and simulate the traffic patterns of legitimate applications, which can be used to manipulate the network behavior of malware traffic. They used a Facebook chat client that directly communicates with the malware to train the GAN and reported promising results in evading a targeted Intrusion Prevention System (IPS).

Yuan et al. [Yua+20] introduced GAPGAN, a framework that can generate adversarial malware examples using GANs that preserve the object's integrity. Their goal was to use the generator network to produce adversarial payloads, which can then be appended into the malware binaries. This approach targets classifiers that accept input in the form of raw binaries, such as MalConv. The authors reported that an adversarial payload can be effectively generated within 20 milliseconds, reaching a 100 % evasion rate. Moreover, the attack is performed using payloads that represent only 2.5 % of the data length.

Visaggio et al. [Vis+21] trained multiple neural networks using different sets of PE features, which were randomly sampled from a training dataset. They selected the 10 best performers in terms of accuracy to build 10 parallel GANs, in which 10 generators were trained using the chosen classifiers as discriminators. In this scenario, the attack uses gradient descent with the maximum mean discrepancy as a distance function to observe the similarity of sample distribution in the training stage. In their study, the authors performed a gray-box attack, as the feature space is known for the adversary, and reported a success rate of around 98 %.

In summary, similar to gradient attacks (§ 2.3.1), GAN-based approaches leverage the feature domain with white-box scenarios, in which the adversary can retrieve internal data from the model. This is why the evasion rates achieved are often close to perfect (100 %).

2.3.3 Reinforcement Learning

In 2018, Anderson et al. [And+18] introduced the first attack generating adversarial malware examples using Reinforcement Learning (RL) for Windows PE files. This approach is based on the OpenAI gym [Bro+16] environment, in which the agent is trained with a policy that aims to maximize a simple reward function that returns *zero* if the modified malware remains detected and 10 when an adversarial example is found. In this scenario, for every episode, the agent receives a budget of 10 turns, which are the number of perturbations injected into the PE file until the next episode starts. However, although the authors reported an evasion rate reaching up to 24 % for specific scenarios (i.e., depending on the type of malware), no integrity verification was performed, since they argued that by implementing semantic-preserving perturbations functionality is expected. Nevertheless, they later acknowledged that multiple files exhibited integrity issues regardless of the type of transformation.

Fang et al. [Fan+19] implemented a value-based double deep Q-network that was originally proposed [VGS16] to tackle the overestimation of action values caused by basic Q-learning algorithms. With this approach, the authors reported an evasion rate of 46.56 %. To achieve such an evasion rate, they limited the feature space from 2, 351 to 513 dimensions representing around 20 % of the input space. Additionally, they reduced the action space to only four transformations. Unlike the previous work [And+18] that they built upon, the authors implemented integrity verification using both a controlled environment [Cla+21] and a static analysis. However, to achieve the reported performance, they needed to inject a sequence of 80 transformations to significantly modify the structure of the PE and, hence, increase the likelihood of nonfunctional adversarial examples.

According to Y. Fang et al. [Fan+20], previous approaches rely on the same feature space for both the agent and the classifier, resembling gray-box instead of black-box settings. Therefore, they introduced a new classifier trained on a modified feature space of 2, 478 dimensions to avoid conflicts with the input space used by the agent. However, unless in the case of implementing featureless approaches, static malware classification usually relies on domain knowledge, which is why it is expected to have shared portions of the feature space between the models and the agents. To mitigate the random components inherent to certain transformations throughout the injection process, such as import functions or compression with random bit rates, the authors created separate transformations for each case by extending the action space to 218. Similar to the strategy explored by Fang et al., [Fan+19], this approach also required extensive sequences reaching up to 100 perturbations to reach successful evasion.

Zhang et al. [ZLC20] introduced an attack that uses RL to train an agent that sequentially injects semantic *nops* into Control Flow Graphs (CFGs). They claimed that this process does not affect the semantics of the malware object, and they implemented CFGs as a connection between code obfuscation and adversarial examples. Then, they performed an attack targeting two kinds of Graph Neural Networks (GNNs) that are trained to detect malware on the basis of CFG-based features using a dataset that includes five different types of Windows malware (e.g., worms and ransomware) and a set of benign software. As a result, the authors reported perfect evasion for the attack strategy, highlighting improvements over baseline methods, such as random, accumulative, and gradient-based instruction insertion attacks, which preserve semantics.

2.3.4   Genetic Programming

In addition to RL strategies, Genetic Programming (GP) has been implemented to generate adversarial examples across multiple domains. Genetic algorithms were introduced by John Koza [Koz90] and are inspired by natural selection processes resembling Darwinian evolution. More specifically, GP explores the algorithmic search space and can be broadly classified as a subset of ML algorithms. Its goal is to drive a population of objects using genetic-like operations, such as *mutation, crossover*, and *fitness* to the next generations until it converges into a solution [KK92].

Noreen et al. [Nor+09] introduced a framework to further develop malicious software that consists of three modules, a code analyzer that creates an abstract representation of the object, an algorithm that processes the representation using genetic operators, and a code generator that converts back high-level representations to machine code. They focused their experiments on further developing a known computer worm, dubbed Bagle [Roz05], and were able to evade mostly signature-based classifiers instead of ML-based models.

Similarly, GP has been useful in finding adversarial examples in the mobile malware domain evaluating both research and commercial classifiers [AS15; RCJ13; ZLL12]. Calleja et al. [Cal+18] proposed using genetic algorithms to manipulate malware labeling. By modifying a single feature, they were able to force an Android classifier to detect some malware as a target family 28 out of 29 times. They finally proposed an improved version of the classifier to tackle these shortcomings. Note that while this approach yielded successful results, the goal was ultimately to fool the model to classify the input object using a wrong malware family instead of evading the classifier altogether.

Adversarial examples with GP are not solely generated using malicious applications, since evolutionary approaches can also be extended to PDF classification. Xu et al. [XQE16] performed a case study on how to modify malicious PDF files to be misclassified as benign by a model. They implemented a stochastic approach inspired from genetic algorithms with *mutation operators* to modify the structure of PDF malware. In this scenario, each operator can be an insertion, deletion, or replacement. The results obtained revealed perfect evasion when automated attacks targeting two PDF classifiers, PDFrate [SS12] and Hidost [ŠL13], were used.

Choi et al. [Cho+19] presented an adaptive approach using GP that relies on open-source input files. They evaluated two scenarios using malware based on Python and C. In this scenario, the attack module processes the source code of the object parsing variables and functions. The attack includes two groups of transformations: nonbehavioral and behavioral. In the former, the changes do not impact the execution of the object, whereas in the latter, the changes add, delete, or swap lines of code, affecting the execution flow. Hence, the authors implemented a verifier using Python and C compilers to ensure that the adversarial example remains functional. However, it is important to highlight that the attack targeted TLSH [OCC13] and Variant [UZ15], legacy malware frameworks based on similarity detection rather than malicious features.

Demetrio et al. [Dem+21] proposed an approach using PE files that is built upon earlier studies [And+18; AR18; LSR19b; LSR19a]. They implemented two transformations: creating a new section and appending bytes at the end of the file, a portion known as overlay. They evaluated their approach by targeting two models, the previously introduced MalConv and a pretrained Light Gradient Boosting Machine (LightGBM) [Ke+17] with the EMBER dataset [AR18]. However, although the results reported seem to show successful evasion of the classifiers, largely relying on only two transformations may compromise the strength of the attack. For instance, the bytes appended at the end of the file may potentially be flagged by preanalysis. Moreover, the MalConv model [SCJ19] does not process positional information properly from input features and is, therefore, vulnerable to attacks that depend on appending data.

Jin et al. [Jin+21] proposed a genetic algorithm by adding two new transformations to the set introduced by Anderson et al. [And+18] and categorized them into behavioral and nonbehavioral changes. They reported improvements of 18 % and 25 % compared to the state of the art [LSR19a], including a 73 % evasion rate against specific targets. However, in their scenario, the False Negative Rate (FNR) seems to be based on all commercial classifiers extracted by VirusTotal [Vir21b] instead of using the same model. Despite being interesting, generalizing over too many targets may lead to suboptimal solutions, since the GP algorithm needs to optimize for multiple classifiers. Furthermore, the modified file's behavior is checked using a sequence of API calls with 30 files. Therefore, no systematic integrity verification stage is conducted for each adversarial example.

Yuste et al. [YPT22] introduced an approach for exploiting and eventually widening the space in between sections of the PE stored on disk, which is later not loaded to the memory but is still evaluated by the models. The process was divided into two stages: an exploratory stage, whose goal was to build the initial adversarial example, and an optimization stage, whose goal was to improve the content of the block to maximize evasion. From the results obtained, the authors reported a success rate of 97.99 % against MalConv, which is an improvement over earlier studies [Dem+21]. However, they highlighted that the previous limitations [Krč+18] of the target model may also apply to this approach. Therefore, as future work, they recommended testing their attack against other models.

2.3.5   Universal Adversarial Perturbations

Universal Adversarial Perturbations (UAPs) are object-agnostic sets of transformations that can cause neural networks to misclassify a wide range of input objects. Moosavi-Dezfooli et al. [Moo+17] demonstrated that a single perturbation, almost indiscernible to the human eye, can force a neural network to modify a correctly predicted label of an image with a high probability. Similarly, the same approach can be extended to domains that are beyond image processing.

UAP attacks have been implemented across several domains, including image processing   [Ath+18; KO18; Bro+17; Co+19; Mop+18], audio classification [Abd+19; Nee+19], and perceptual ad-blocking [Tra+19], in which realizable attacks can be generated.

Hou et al. [Hou+20] explored the use of UAPs in the feature space in the context of Android malware. They extracted API and hardware information to generate a binary vector that can be modified to target the classifier with PK. They attacked three models and reported full evasion against the DREBIN model [Arp+14] after nine transformations. The remaining two target models required up to 65 and 100 perturbations, respectively, to achieve an adversarial evasion rate close to 100 %.

Recently, Rashid et al. [RS22] attempted to detect universal adversarial examples using Moving Target Defense (MTD) capabilities, which have been found to mitigate network-level threats [Pos+20]. However, although this was a novel approach, it seems to focus mainly on feature-space adversarial examples rather than on real-world UAP attacks.

Other researchers have implemented additional strategies beyond those addressed in § 2.3.1–2.3.5. For example, Lucas et al. [Luc+21] proposed an approach using software diversification to generate adversarial examples that leverage stochastic optimization techniques to guide malware manipulations. The attack targeted two deep neural networks, MalConv and Krčál et al., [Krč+18], which have been trained using a raw byte-level representation of malware objects. The authors reported near-perfect evasion using functionality-preserving perturbations, yet they verified the integrity only for a handful of files at the end and not throughout the process. With this approach, they reported near-perfect evasion scores. Therefore, they suggested that malware detection should also leverage techniques beyond ML.

2.4 Summary

From the literature research presented in Table 2.1, we have identified gaps that need to be filled in order to create a comprehensive framework that includes several aspects from the AML field in the context of malware. In other words, the proposed solution should support byte-level perturbations in the problem space along with a verification stage in the form of an integrity test. The underlying idea is to recreate a scenario in which classifiers can be evaluated using attacks similar to real-world ones. To maximize the efficiency of the proposed solution, such a framework should consider multiple types of attacks, including most of the strategies presented in the literature. This will help provide consistent benchmarks, which are also missing in the field. Our goal is to develop a solution that focuses on supporting PE files, especially EXE. This is because not only is Windows malware the most predominant threat [AV-20], but also it seems to be underexplored in the field compared to PDF and Android Package Kits (APKs), since both are, respectively, flexible to perturbations and open-source and, therefore, less complex to manipulate.

Table 2.1 A literature review of adversarial attacks in the malware context. The symbols are defined as follows: ✓ = include, (✓) = partially include, ✗ = do not include, and ✓+ = multiple attacks types, including Stochastic, GP, RL, UAP, GAN, and Gradient Optimization. The integrity test checks whether the functionality of the object is verified, whereas PS refers to attacks realizable in the problem space

RELATED WORK	BYTE PERTS.	INTEGRITY TEST	ATTACK TYPE	PS REAL.	FILE TYPE	FEAT. ANALYSIS
Biggio et al., 2013	✗	✗	Gradient	✗	PDF	(✓)
Xu et al., 2016	✗	✗	GP	✓	PDF	✓
Hu and Tan, 2017	(✓)	✗	GAN	✗	EXE	✗
Grosse et al., 2017	✗	✗	Gradient	✗	APK	(✓)
Anderson et al., 2018	✓	✗	RL	✓	EXE	✗
Kreuk et al., 2018	(✓)	(✓)	FGSM	✓	EXE	(✓)
Al-Dujaili et al., 2018	✗	✗	dlrFGSM	✗	EXE	(✓)
Kolosnjaji et al., 2018	(✓)	✗	Gradient	✗	EXE	(✓)
Pierazzi et al., 2019	✗	(✓)	Gradient	✓	APK	(✓)
Choi et al., 2019	(✓)	✓	GP	✓	EXE	(✓)
Fang et al., 2019	✓	(✓)	RL	✓	EXE	✗
Y. Fang et al., 2020	✓	✓	RL	✓	EXE	✗
Demetrio et al., 2020	✓	✗	GP	✓	EXE	(✓)
Hou et al., 2020	✗	✗	UAP	✗	APK	✗
Zhang et al., 2020	✓	✗	RL	✗	EXE	✗
Yuan et al., 2020	(✓)	✗	GAN	✓	EXE	✗
Jin et al., 2021	✓	(✓)	GP	✓	EXE	✓
Lucas et al., 2021	✓	(✓)	Stochastic	✓	EXE	✗
Visaggio et al., 2021	✓	✗	GAN	✗	EXE	✓
Yuste et al., 2022	✓	✗	GP	✓	EXE	(✓)
Framework	✓	✓	✓+	✓	**EXE**	✓

Part II
Framework for Adversarial Malware Evaluation

FAME

Unity is strength. Synergy is might.
—Matshona Dhliwayo

On the basis of the literature evaluated in Chapter 2 and the limitations outlined in Table 2.1, we introduce our Framework for Adversarial Malware Evaluation (FAME) [LR22], which can be observed in Fig. 1.1 under a homonymous name. We define the notation and threat model for our research by describing the adversary's knowledge, objectives, and capabilities. Since the requirements vary depending on the attack settings, they will be presented individually in the subsequent modules of Part III.

3.1   Notation

We start by introducing the notation that will be used throughout our framework, following the work of Biggio and Roli [BR18]. In general, an input object can be defined in two categories: *feature space* and *problem space*. The feature space refers to the embedded representation processed by the model, whereas the problem space represents real objects (i.e., software binaries). Our framework supports attack strategies in both domain spaces. Because of the differences among feature representations and real objects, we define the constraints for each domain that need to be addressed when producing adversarial examples.

The feature and label spaces are, respectively, denoted by $\mathcal{X}$, $\mathcal{Y}$, whereas the problem space is represented by $\mathcal{Z}$. For each input object $z \in \mathcal{Z}$, a corresponding ground-truth label $y \in \mathcal{Y}$ is associated. A classifier $g : \mathcal{X} \longrightarrow \mathcal{Y}$ predicts a label $\hat{y} = g(x)$ for the given input object, and a feature mapping function $\varphi : \mathcal{Z} \longrightarrow \mathcal{X} \subseteq$

R. Labaca-Castro, *Machine Learning under Malware Attack*, https://doi.org/10.1007/978-3-658-40442-0_3

$\mathbb{R}^n$ is required by the classifier to process the input object. Such a function is not differentiable and not invertible, which makes the process of finding problem-space attacks using gradient methods nontrivial, given that it does not allow mapping features back to the problem space. Hence, a number of constraints are required to successfully generate adversarial examples in the problem space.

3.1.1   Feature-Space Attacks

To evaluate malware classifiers, we create a scenario in which a potential adversary generates attacks by modifying the input object (i.e., malware file) until it is no longer detectable by the model. This means that the the adversary aims to convert an object $x \in \mathcal{X}$ into $x' \in \mathcal{X}$, in which $g(x') = t \in \mathcal{Y}$ where $t \neq y$, thereby forcing the model to assign the wrong class t to an object x'. In the feature space, the goal is to trick the model into assigning a benign label to a malicious object represented in the embedded domain.

3.1.2   Feature-Space Constraints

While feature-space attacks do not produce real-world objects, they still need to meet a group of restrictions to resemble viable objects. Thus, a set of constraints Ω specify the transformations occurring in the feature space, which can include for instance lower lb and upper ub bounds such that $\delta_{lb} \preceq \delta \preceq \delta_{ub}$. They can also limit the number of features from the input object affected by transformations.

3.1.3   Problem-Space Attacks

In the problem space, attacks must be carefully constructed to generate real objects [Pie+20]. The purpose of the adversary is to generate a sequence $\mathbf{T} : T_n \circ T_{n-1} \circ \ldots \circ T_1$ for which every transformation $T : \mathcal{Z} \longrightarrow \mathcal{Z}$ converts z into an object $z' \in \mathcal{Z}$ such that $g(\mathbf{T}(z)) = t \in \mathcal{Y}$ where $t \neq y$, in which all problem-space constraints are correctly satisfied. This process leads to the generation of adversarial malware examples in such a way that the model wrongly labels them as benign.

3.1.4   Problem-Space Constraints

Since attacks need to be carefully executed in the problem space to ensure that the integrity of the object is preserved, additional constraints are required. Generally, malware perturbations, especially in the context of PE files, largely rely on adding or eventually editing information, since removing data may compromise the integrity of the object. Adding functions to the IAT and manipulating the signature or debug information of the binary are examples of valid perturbations. Hence, an integrity verification stage, defined as $v : \mathcal{Z} \longrightarrow \mathcal{Z}$, is included to ensure that the transformations in question do not compromise the integrity of the adversarial examples generated.

3.2 Threat Model

In this section, we define the threat model while considering multiple aspects from the adversary's perspective during *evasion attacks*, in which the attackers aim to evade the model in the *test-time*. We start by defining the goals that will drive the attack modules. Generally, adversary knowledge is represented using two levels: PK, which provides the adversary with full access to the model (i.e., white box) and settings, and LK, which provides only restricted information from the classifier, such as hard labels.

3.2.1 Adversary Objectives

In general, the goal of the adversary is to increase the FNR of the model by identifying optimal sequences of transformations that when injected into an input object drive the model to predict a benign label for a given malware example. These evasion attacks are conducted during test time only [Big+13] and, unlike poisoning attacks [BNL12], do not rely on compromising the data during training time.

3.2.2 Adversary Knowledge

In this work, we initially assume that the adversary has LK about the target to reproduce and evaluate realizable attacks. Later, we explore further scenarios in which the adversary has PK to assess the impact of feature domain attacks in the context of malware. Following the line of work conducted by Biggio et al. [BR18] and shared by Carlini et al. [Car+19], we define the training data knowledge by $\mathcal{D}$ with feature set $\mathcal{X}$, algorithm g, and hyperparameters w.

Perfect Knowledge (PK)

This scenario, $\theta_{PK} = \{\mathcal{D}, \mathcal{X}, g, \boldsymbol{w}\}$, is only adopted for feature-space attacks in our framework. Here, the adversary has complete knowledge of the learner and hyperparameters. Otherwise, the target model can be unconditionally queried to retrieve the soft labels. This setting facilitates strong attacks that implement gradient optimization to produce optimal adversarial examples.

Limited Knowledge (LK)

Under LK settings, $\theta_{LK} = \{\hat{\mathcal{D}}\}$,the adversary can send the model unlimited queries and receive in return the predicted hard labels. In this scenario, the optimal attack is calculated by maximizing the optimization functions that include multiple parameters, such as similarity, distance from the first generation, and the predicted class. For universal adversarial attacks, the model returns soft labels to every query. Thus, the input data distribution can be approximated to select the most effective sequence of transformations to apply in the attack. In case the adversary has additional knowledge about the feature set and the learning algorithm $\theta_{LK} = \{\hat{\mathcal{D}}, \hat{\mathcal{X}}, \hat{g}, \hat{w}\}$, then a surrogate model (i.e., a substitute model trained to fit the target classifier) can be queried.

3.2.3   Adversary Capabilities

Since access to the input object is unrestricted for adversaries, then they have an unlimited time to craft domain-specific byte-level perturbations. However, since Windows malware is mostly not open-source, to keep the attack scenario as realistic as possible, we assume that the adversary has no access to the source code of the input objects.

Overall, the size of the transformations is not limited in terms of L_p norm [Pie+20]. However, it is recommended to use smaller over larger [Fan+19; Fan+20] sequences of transformations, since shorter sequences seem to be more appropriate for the prevention of nonfunctional PE files.

3.3   Experimental Settings

In this section, we describe the experimental settings under which the classifiers are tested to evaluate their robustness against input-specific and universal attacks that produce adversarial malware examples. Initially, we tested adversarial objects against a wide range of classifiers using an aggregator platform [Vir21b]. However,

because of availability constraints and reproducibility reasons, we locally implemented research models trained with publicly available datasets. Moreover, byte-level
perturbations and integrity verification stages are available as open-source software.

3.3.1 Target Model

The targeted model is a highly efficient variation of a Gradient Boosted Decision
Tree (GBDT) known as LightGBM [Ke+17], which has been trained with 600,000
benign and malicious PE files and has achieved an AUC-ROC score of 0.993. For
training purposes, the EMBER dataset [AR18] was used, which is currently widely
implemented in the literature [AE20; Fan+19; Sev+21; Vis+21]. Following previous
implementations [And+18], the confidence rate of the model was set to 0.9. Above
this threshold, the object is classified as malicious, whereas below this threshold,
the object is classified as benign, hence an adversarial example. However, this information is not provided to the adversary and can vary depending on the configuration
of the model.

A GBDT, proposed by Anderson et al. [And+18], was implemented in early
experiments and will be referenced. This GBDT model was trained using 100,000
malicious and benign Windows PE files and achieved an AUC-ROC score of 0.993
with a 90 % True Positive Rate (TPR) and a 1 % False Positive Rate (FPR).

3.3.2 Datasets

All malware classifiers used as target models during our experiments were trained
using an open-source dataset. In this section, we also describe an additional dataset
from which problem-space files are sampled to be used as input objects in the
attack modules, as well as a pair of subsets specifically used to search for universal
perturbations.

EMBER

The Windows malware dataset introduced by Anderson et al. [AR18] contains the
feature space of 800,000 PE files divided into 400,000 benign and 400,000 malicious
files. Overall, 96 % of the data was collected between January and December 2017,
whereas the remaining 4 % of the files anticipate these dates. The files were assigned
to the *malicious* class if more than 40 engines detect them in VirusTotal and *benign*
if no detection is reported.

Unlike similar malware datasets [Al-+18; Arp+14], the security domain includes continuous features that are categorized as follows: parsed features, which include header information; format-agnostic histograms, such as entropy and byte-value histograms; and printable strings, which count the average length and URL frequency.

FAME

For the problem-space domain, we collected 6,880 Windows PE files from the VirusShare platform [Vir21a]. This dataset is sampled as an input object when a module is triggered by the framework in order to generate an adversarial example. The size of each file is constrained to a lower boundary of 100 kilobytes to ensure that the transformations do not dominate the structure of the binary.

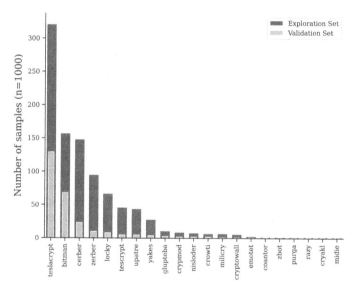

Figure 3.1 Adversarial examples from *exploration* and *validation* sets grouped by malware family. It can be observed that the attacks using the toolbox of byte-level perturbations are applied to multiple input objects regardless of the malware nature to avoid bias toward a particular family

UAP Exploration & Validation

Two subsets from the $FAME$ dataset were created as $exploration$ (n=1000) and $validation$ (n=100) to be used by certain modules in the framework (i.e., during

UAP generation). To avoid bias during the UAP search, for both the *exploration* and *validation* sets, we sampled files from heterogeneous malware families, as shown in Fig. 3.1. These families represented a wide distribution of malware types consisting of more than 20 families, including well-known ransomware, such as *Teslacrypt* and *Locky*, and Trojan downloaders like *Upatre*. We therefore expect our approach to be well generalizable regardless of the malware nature.

3.3.3 Byte-Level Perturbations

To generate effective and realizable adversarial malware examples, we added sequences of transformations to an input object by querying a toolbox that contains a set of byte-level perturbations that aim to preserve the functionality of the file. These transformations act as noise that aims to change only the structure of the object but not its functionality.

Hence, the attacks used realizable byte-level perturbations that were introduced by Anderson et al. [And+18] and shared by other studies in the literature [HT17]. However, two of the transformations that were initially proposed in the set, namely, *identity* and *create_new_entry_point*, were removed because of some technical problems that impacted the adversarial examples. Given that the input objects were closed-source binaries, the static features of the file were extracted with LIEF [Tho18], an external cross-platform library that can manipulate PE files.

Perturbations are classified into three groups: *i) inclusion*, which refers to transformations that append bytes or add new functions to the file; *ii) modification*, which groups changes that manipulate information in the header or rename sections, and *iii) compression* that leads to the packing or unpacking of PE files with the Ultimate Packer for eXecutables (UPX) [MLJ20] using randomly defined compression rates.

A set of byte-level perturbations, as shown in Table 3.1, are described as follows. (i) *overlay_append* appends a sequence of random bytes at the overlay of the PE, a portion also known as overlay. (ii) *imports_append* samples a function from a list of DLL imports and adds it to the IAT. (iii) *section_rename* manipulates sections by replacing them with the name of benign sections. (iv) *section_add* adds a section to the section table that will not be used. (v) *section_append* appends random bytes at the end of the sections instead of adding a section. (vi) *upx_pack*: compresses the file with UPX at a random compression rate. (vii) *upx_unpack*: decompress the PE using UPX. (viii) *remove_signature* breaks the signature from *CertificateTable* in the data directories' optional header. (ix) *remove_debug*: manipulates the *Debug* information located in the data directories' optional header. Finally, (x)

Table 3.1 Set of realizable byte-level perturbations that are added to Windows PE files to generate adversarial malware examples against a target model

t_i	DESCRIPTION
t_0	Overlay Append
t_1	Imports Append
t_2	Rename Section
t_3	Add Section
t_4	Section Append
t_5	Remove Signature
t_6	Remove Debug
t_7	UPX Pack
t_8	UPX Unpack
t_9	Break Optional Header

break_optional_header_checksum replaces the *Checksum* with zero in the optional header.

Generally, while adding a list of transformations needs to be carefully automated, removing parts of the software binary may require extra resources and potentially a manual review to be correctly executed. Thus, perturbations typically add or change information in specific fields, mostly in the header of the PE file, but do not explicitly remove or delete sections to avoid potentially corrupting the file.

3.3.4   Integrity Verification

In this section, we describe the implementation of the integrity verification stage using a controlled environment to ensure that the experiments are conducted successfully under secure settings.

Despite byte-level perturbations being designed to preserve semantics, we have empirically determined that the integrity of the adversarial examples is often compromised after a number of perturbations are injected. This behavior seems to be linked to the heterogeneous nature of PE files and the obfuscation present, which has also been flagged by other authors [And+18]. In this context, building the IAT after the injection of manipulations may require extra patches to ensure that the object remains valid, which explains how certain perturbations, such as *imports_append*, can lead to problematic adversarial examples. Therefore, it is not trivial to map such byte-level perturbations with their impact on the files making an integrity verification step paramount to ensure PE files remain functional. Moreover, improving the functionality rates can be an expensive endeavor. Thus, we significantly optimized our environment and reduced the time required to analyze an adversarial example to ~25 seconds on average. When properly implemented, a verification step generates realizable attacks at a reasonable cost.

The integrity verification stage is built as an isolated environment to follow safety precautions when running adversarial malware examples. The environment includes an instance of the Cuckoo Sandbox [Cla+21], which is an open-source comprehensive malware analysis platform that is widely distributed across the research community. This environment provides a number of features beyond what is needed for FAME and is, hence, implemented in a more restricted manner to reduce the time of processing modules.

3.4   Summary

In this section, we formally introduced FAME and described its characteristics, including the adversary's goals and knowledge, along with malware classifiers and the respective feature set for training. We also defined an additional problem-space dataset that is used to sample input objects. To ensure that attacks in the problem space produce realizable PE files, we implemented an integrity verification stage that checks the functionality of the adversarial malware examples by executing them inside a contained environment. Further approaches that seek to ensure that the file remains valid, such as static analysis and execution workflows, may be too expensive to implement and complex to automate and were, therefore, excluded from this framework.

Part III
Problem-Space Attacks

Stochastic Method

4

Life, in the end, is nothing more than a sum of minutiae that, through random circum-stances, become transcendental events.

—Federico Andahazi

As introduced in Part II, FAME is built in a modular fashion to allow increased compatibility with further extensions. Each module aims to either attack the target model and identify the weaknesses of static malware classifiers (Parts III and IV) or enhance the model's resilience against adversarial attacks (Part V).

In this section, we start by describing attacks and evaluating their performance against a malware classifier. The goal is to generate adversarial malware examples using multiple strategies to optimize the identification of weaknesses in the models. In Chapters 4–7, we assume that the adversary has only LK about the target model. Thus, our goal here is to explore multiple strategies to find the most effective attacks by only allowing the attacker to query hard labels. Note that for universal adversarial attacks, the model returns soft labels to effectively allow for comparisons when performing a greedy search. In Chapters 8 and 9, the adversary attacks white-box models with PK (§ 3.2.2).

In Chapter 10, we compare all approaches and explore their strengths and weaknesses. However, because of the idiosyncratic nature of each strategy (e.g., RL agents focus on maximizing the reward function, whereas GP approaches aim to converge toward an optimal solution), their evaluation settings may differ depending on the strategy implemented. Therefore, we perform an additional unified evaluation with freshly defined assumptions to allow diverse approaches sharing the same environmental settings to be compared.

We now introduce the module for Automatic Random Manipulations to Evade Detection (ARMED) [LSR19a], which is depicted in Fig. 1.1 as the first problem-space attack. The goal of ARMED is to automate the pipeline for finding vulnerabilities in the models by identifying adversarial examples in the context of malware while ensuring that the output objects are realizable. The process relies on randomly searching for the best sequences of transformations until a given threshold of successful adversarial examples is met.

4.1   Requirements

To ensure that end-to-end attacks can generate valid adversarial malware examples without manual intervention, we considered four fundamental requirements.

– *Automation*: Every adversarial example is generated automatically. In other words, after an input object is provided, ARMED searches for the right sequence of transformations in which the PE file is no longer flagged as malicious by the target model.
– *Platform*: Byte-level perturbations are carefully designed to work with Windows PE files and are, hence, the focus of the module. However, the architecture of ARMED is built to be platform-agnostic to maintain future compatibility with alternative platforms.
– *Size*: While we do not *per se* restrict the length of the transformation sequences[1], we empirically define a lower boundary of 100 kilobytes for the size of the PE files. To this end, we ensure that small files are not drastically impacted by the adversarial perturbations.
– *Plausibility*: Because of the problem-space constraints, we verify the integrity of every adversarial malware example generated before shifting the attention to bypassing the classifier.

4.2   Methodology

Fig. 4.1 shows the workflow used to generate adversarial examples automatically by maximizing the evasion rate. The process starts with the generation of a random sequence of transformations that are then injected into the input object during the manipulation stage. Then, after the modified object is created, it is submitted to the integrity verification stage to ensure that its structure remains valid before the classifier assigns it a label.

---

[1] We specify sequences or chains of transformations as vectors $|\mathbf{T}| = n$ represented as $(t_1, t_2, ..., t_n)$, without an upper boundary.

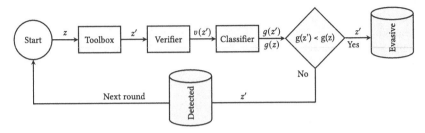

Figure 4.1 ARMED workflow. The input object z receives a randomly chosen sequence of transformations from the toolbox, and the verifier ensures that the object remains valid before the classifier returns a detection rate. If the level of evasiveness increases (classification rates are minimized), then an adversarial example z' is produced

4.2.1 Toolbox

The toolbox consists of a set of byte-level perturbations that are grouped in sequences of different variations to be injected into the input object to produce adversarial examples. While there are no boundaries attached to the length of each sequence, multiple values can be explored to evaluate the relationship between the number of transformations and the effectiveness of the attack. Moreover, the attack is designed to be agnostic to the toolbox, which can be extended, reduced, or replaced without impacting the workflow.

4.2.2 Verifier

During the verification stage, a safe environment is used to ensure that the adversarial example meets the integrity criteria to allow executability. Additional features are provided within the environment, such as files dropped and traffic dumps, to evaluate the level of maliciousness. However, these features are not required for the integrity verification stage and are, therefore, deactivated by default.

4.2.3 Classifier

The model is a malware classifier in charge of returning a prediction (malicious or benign) to the object analyzed. An extensive evaluation was performed using 65 commercial classifiers from the VirusTotal [Vir21b] aggregator platform[2]. Note that if a single target model is used instead, then $g(z') < v$ where v is the threshold below which the model classifies the object as malicious.

Let us now break down the process of generating adversarial examples with ARMED into three steps. In *Step 1* the *toolbox* receives an input object z and injects a single randomly calculated sequence of transformations (**T**). Once the adversarial example z' is retrieved from the *toolbox* the *verifier* in *Step 2* assesses its integrity. In case of corruption (nonfunctional), the module loops back to the *toolbox* and requests a new sequence of perturbations to generate a fresh adversarial example. In *Step 3* the *classifier* predicts labels for the input object z and the adversarial example z' to be able to compare scores. While the input objects are expected to be malicious, the classification rates across malware classifiers can vary and should, hence, be benchmarked before the success of the attack is evaluated. If the adversarial example is successful, then the classifier returns a label that differs from the original input object.

4.3 Evaluation

In this section, we evaluate ARMED in terms of two metrics, namely, *integrity* and *evasiveness*, since both aspects are required to generate valid and effective adversarial examples.

4.3.1 Integrity

As described in § 4.2, to ascertain whether the adversarial examples remain functional, we generate 1,265 adversarial examples with random sequences of transformations, which are assessed during the integrity verification stage.

Fig. 4.2, shows cumulative results as a summary of this evaluation. For every sequence containing 2–25 transformations, we generate as many adversarial examples as needed until we find 10 valid objects (light gray in the figure), hence functional files. We start the evaluation with chains of size two, since empirical experiments have shown that adversarial malware examples with single transformations may

---

[2] We are aware that the classifiers available in aggregators often suffer from overparametrization, which can lead to higher classification rates at the expense of false positives [Vir21c]. Thus, we suggest retrieving labels from these platforms carefully, since this can yield different results compared to local implementations, depending on how sensitive the models are implemented.

end up very similar to the input object and can lead to unusual collisions. Likewise, we identify a clear trend after 20 and, therefore, decide to evaluate until 25, since we assume that the cost would discourage the adversaries from pursuing realizable attacks. All adversarial examples that have been created in excess until the goal is met are shown in dark gray.

For example, almost 30 adversarial examples were generated for $|\mathbf{T}| = 2$ to obtain 10 valid malware examples. Similarly, over 40 files were required when the sequence consisted of 10 transformations. Overall, the figure shows a clear trend when the number of transformations injected increases, since significantly more attempts are needed to obtain only 10 valid files. For $|\mathbf{T}| = 20$ onwards, we need to generate more than 100 adversarial examples, which reveals how expensive it is to generate large successful vectors. Additionally, all cases in which $|\mathbf{T}| > 20$ achieve values significantly above the average (red line) and are, thus, less appealing to pursue in terms of integrity levels.

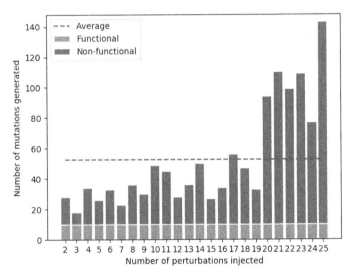

Figure 4.2 Distribution of adversarial examples based on the input object z_a (n = 1,265). Each column indicates the number of attempts required to generate 10 functional adversarial examples (light gray). The cumulative values of nonfunctional files (dark gray) show how the cost to generate valid adversarial examples increases substantially when longer sequences are added to the input object

4.3.2   Evasiveness

For the next metric, evasiveness, we evaluate the adversarial examples generated from an input object previously labeled as malicious by 46 out of 65 commercial malware classifiers.

To assess the impact of the sequence on the classification rate, we sampled an input object z_b and generated 240 adversarial examples with sequences of lengths ranging from 2 to 25. We limited the evaluation to 25 to follow the experiments in § 4.3.1. Given that the detection rates do not seem to improve afterward, this seems to be a reasonable threshold. We produced 10 adversarial examples for each case.

The results are shown in Fig. 4.3, in which the blue line shows the average of the classification rates. Unlike in Fig. 4.2, there is no clear trend between the number of transformations injected and misclassification. In contrast, we can observe that larger sequences are not necessarily better at evading classifiers.

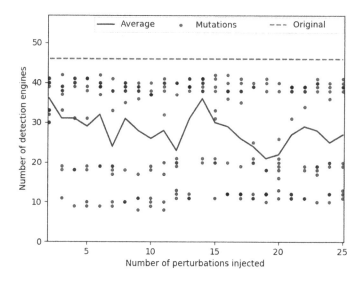

Figure 4.3 Distribution of 240 adversarial examples produced from the input object z_b. The figure shows that for each new adversarial example generated, the classification rate is reduced. While there seems to be a slight trend on average toward higher evasion rates with longer sequences (e.g., 12 and 20), the most evasive adversarial examples are found by adding around 10 transformations to the object, as observed from the points located at the bottom left

For instance, $|\mathbf{T}| = 12$ and $|\mathbf{T}| = 19$ report the best averages, whereas the single most successful adversarial examples received $|\mathbf{T}| = 9$ and $|\mathbf{T}| = 11$ transformations, respectively, as indicated by the gray dots at the bottom of the figure.

4.3.3 High-Dimensional Sequences

We evaluated adversarial examples with high-dimensional attacks to assess the impact on the classification rates. While we were unable to generate valid examples for all cases (z'_n) derived from z_0, we compared the performance among the best adversarial examples across thousands of files generated with diverse lengths of sequences of perturbations.

As shown in Table 4.1, longer sequences do not necessarily lead to lower evasion rates. In fact, we obtained the same evasion rate for $|\mathbf{T}| = 5$ and $|\mathbf{T}| = 500$. One reason for this may be that the more the perturbations injected, the more likely the classifier is to flag suspicious behavior. Moreover, perfect evasion $(g(z') = 0)$ has not been reported for any of the adversarial examples. This is because a minor subset of the classifiers always flagged the binary as malicious, which indicates a high sensitivity with a potentially affected specificity.

Table 4.1 Selection of highly evasive adversarial examples based on the classification rate. The goal is to identify sequences of transformations that maximize the number of classifiers evaded. As can be observed, longer sequences do not necessarily have a significant impact on evasion, since the evasion rates remain similar

FILE	LENGTH	EVASION	VALID
z_0	–	**67.69%**	✓
z'_1	5	92.30%	✓
z'_2	50	93.84%	✗
z'_3	500	92.30%	✗

On average, each adversarial example required approximately five minutes to be completely analyzed using the aggregator's platform. This is because of the significant delays experienced while connecting with the API of cloud-based services. Our experiments indicate that the processing time can be reduced by approximately 90 % if we implement a target model locally. However, commercial classifiers are better accessible through integrators, since local deployments may be limited by licensing issues.

4.4   Summary

After evaluating the attack against 65 commercial classifiers, we determined that ARMED can produce highly evasive valid adversarial examples in a reasonable time space. For performance-related reasons, let us now shift our attention toward targeting local ML-based models with state-of-the-art performance.

Despite being successful, ARMED presents intrinsic limitations when scaling the generation of adversarial malware examples. Even after the processing time was substantially reduced during integrity verification, as observed in our experiments, the nondeterministic nature of the module led to lower success rates when the attack was scaled. Although parallel processing may partially address this limitation, the algorithm exploration is too slow and rapidly becomes inefficient for larger attacks.

Another aspect that is worth considering is *future convergence*. Overall, the likelihood that the model will find faster adversarial examples over time is not guaranteed. That is, there is no reliable learning mechanism to account for better convergence rates in the future regardless of how many objects were processed in the past, which significantly limits how fast models can be evaluated.

Genetic Programming

5

Intelligence is based on how efficient a species became at doing the things they need to survive.

—Charles Darwin

In Chapter 4, we explored how adversarial examples in the context of PE malware can be generated automatically, without intervention, using random sequences of transformations. However, as outlined in § 4.3, the integrity verification rates are indirectly proportional to the length of the attack vector and the level of convergence does not increase over time. Moreover, although cloud-based aggregators provide substantial information about prediction labels, they are expensive to implement.

Therefore, optimization algorithms such as GP may be interesting, since they can accelerate the search for adversarial examples in case the adversary cannot retrieve gradient information from the model. Moreover, together with local models, they can significantly reduce the processing time required, as both are consistently more efficient in returning input labels compared to stochastic approaches using cloud-based aggregator platforms.

Given the aforementioned limitations, we introduce the Automatic Intelligent Manipulations to Evade Detection (AIMED) module. AIMED is a genetic approach used to automatically evaluate weaknesses in black-box malware classifiers by searching for optimized sequences of transformations, which can help identify successful adversarial examples more efficiently [LSR19a]. This approach can be observed as the next problem-space attack in Fig. 1.1.

© The Author(s), under exclusive license to Springer Fachmedien Wiesbaden GmbH, part of Springer Nature 2023
R. Labaca-Castro, *Machine Learning under Malware Attack*,
https://doi.org/10.1007/978-3-658-40442-0_5

5.1   Requirements

In this section, we extend the ARMED module requirements in § 4.1 to include two additional stages, which are required when approaching this challenge from a GP perspective.

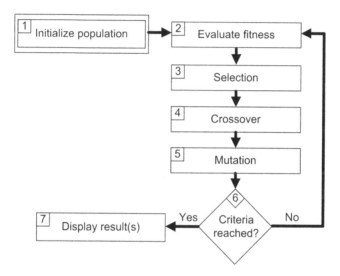

Figure 5.1 Genetic algorithm approach illustrating seven stages. First, the population is initialized using a set of adversarial examples modified with a random sequence of transformations. Then, each generation of members iterates through the different steps until a termination criterion is reached

- *Learning*: Unlike stochastic approaches, attacks using genetic algorithms evolve over time by learning how to improve the sequence of transformations after combining the knowledge of the best attacks. Hence, new generations are expected to be more likely to converge toward an optimal solution.
- *Model*: Local ML-based malware classifiers are assessed using adversarial examples and are responsible for returning the predicted label for each object. Unlike models hosted in cloud-based aggregators, the availability of local classifiers is permanent and the processing time is significantly reduced from minutes to a millisecond range.

5.2   Methodology

In this section, we describe the workflow of AIMED (shown in Fig. 5.2), which consists of a series of stages inspired by the classical evolutionary algorithm approach observed in Fig. 5.1.

Step 1 corresponds to how a *population* gets initialized, which occurs within the *toolbox*, in which each member is generated. Note that during the initialization process, the population is generated by injecting random sequences of transformations into the input object, which initiates the desired population of adversarial malware examples.

In *Step 2*, every member of the generation goes through a fitness evaluation stage, in which a score is assigned. This *fitness score* represents the adversarial readiness of each member to assess the level of eligibility for next generations, and it consists of a function with four parameters, as shown in Fig. 5.2. In this scenario, the *verifier* evaluates the integrity of the new member, whereas the classifier returns the predicted label. Both are extended from ARMED (Chapter 4). *Similarity* refers to the similarity factor between the input object and freshly generated members of a generation, and *Distance* calculates the interval from generation zero, the beginning of the evolution process.

Step 3 corresponds to the selection process inspired from Darwinian evolution, in which the fittest members are selected to start the next generation of offspring, which will potentially have higher chances of evading the model.

For *Step 4*, crossover refers to the selected members mating and creating new offspring. This is the stage at which a new generation arises, assuming that by crossing over the best parents enhanced members will result as offspring.

In *Step 5* a small portion of the population is induced to changes in their attacks by randomly flipping a transformation in their sequence. For example, (t_2, t_1, t_8) becomes (t_2, t_1, t_5) where t_8 is randomly replaced by t_5. These genetic mutations can help avoid suboptimal solutions, in which certain members keep crossing over in loops, thus not increasing the level of diversity.

Step 6 provides the termination criteria. The process continues over generations until a predefined rule is met. In this case, a set of evasive members is identified or a threshold of generations is reached. While in Fig. 5.1 the criteria are checked at the end of the process, the workflow (Fig. 5.2) can be adapted to the problem. Hence, the effectiveness of a member can be checked directly after the fitness evaluation to favor faster convergence in larger populations.

In *Step 7*, the results obtained are presented. The set of evasive members represents the valid adversarial malware examples generated during the process. In case

no effective attacks are found, the set returned is empty and the process restarts with
a new input object.

5.2.1 Genetic Operators

Based on natural evolution, AIMED implements genetic operations to automatically evolve members of the population until a successful solution emerges. These
operators are described as follows:

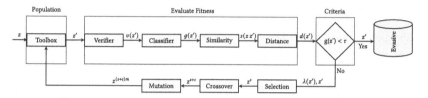

Figure 5.2 AIMED workflow. An input object z is used to generate each adversarial example z', which is then assigned a fitness score. According to such fitness, a set of adversarial
examples are selected to mate and generate the next-generation offspring. Occasional mutations occur to selected members, which may increase the level of evasiveness against the
classifier

- *Selection*: During the selection process, the fittest members (highest fitness score)
  are chosen to breed the next generation of population members.
- *Crossover*: This stage is inspired from the biological process in which two chro-
  mosomes exchange their genetic material to increase the genetic diversity. While
  crossing over improves the level of diversity, nature has shown that a number
  of improvements in species originate from unexpected mutations that can occur,
  for example, during cell division.
- *Mutation*: Genetic mutations are paramount in the evolution process, since they
  increase the level of diversity. However, just like in nature, mutations are not
  always beneficial and can rather be neutral or may even lead to negative results,
  which in the context of malware means compromising the integrity of the adversarial example.

Following the recommendation [KK92] to reproduce only a small portion of the
population to promote diversity, we define the limits to the population selected to
reproduce between 10 %–50 % in accordance with the size of the population.

5.2.2 Fitness Function

To evaluate which members are the best among their generation, we define the fitness function λ in Eq. 5.1 as follows:

$$\lambda(z') = v(z') + g(z') + s(z, z') + d(z') \tag{5.1}$$

where $v(z')$ corresponds to the integrity verification step, which ensures that the members are functional, and $g(z')$ is the detection model that returns the predicted label. In this scenario, the similarity function, $s(z, z')$, calculates the similarity score between two members. Although close relatives (low similarity score) may be interesting to promote in case they are close to become adversarial, this may lead to suboptimal solutions if the members are corrupt. Hence, higher values are promoted as they maximize the level of diversity across the population. Finally, $d(z')$ returns the generation in which members of the population were created, emphasizing the importance of younger offspring.

Therefore, an adversarial member $z_i' \in Z^*$ is calculated by maximizing the fitness function (Eq. 5.2) with an input $z_i \in Z$. The input object is then mapped to an optimal sequence of transformations $\mathbf{T}$ that satisfies $g(\mathbf{T}(z_i')) = t \in \mathcal{Y}$ when detection rate $u < \tau$ where τ is the threshold below which a benign label t is assigned to a malicious input file z':

$$z' = \arg\max \sum_{i=0}^{n} \lambda(z'), \forall z' \in Z \tag{5.2}$$

where n is the number of generations processed until an adversarial member is found. Note that the detection rate is only used to determine the hard label of the binary classification and is not available to the adversary.

5.3 Evaluation

During our experiments, we sampled input objects from the FAME dataset, which has been described in § 3.3.2, and selected over 5,000 of the *fittest* adversarial examples. Overall, the duration of a generation varies between 50 and 250 seconds, depending on the optimization of the environment and the classifiers implemented. We created the initial population by injecting random sequences of transformations with a length of 10 into the input object. Then, fresh members received a score depending on their fitness value, which consists of a function that includes integrity, classification, similarity, and generation. These parameters are chosen to represent the important aspects of an adversarial example and make each one of them more unique among its generation.

Next, integrity is verified with a safe environment (§ 3.3.4), in which each member is evaluated. Classification is performed using a GBDT that yields an AUC-ROC score of 0.993 with 90 % TPR at a 1 % FPR. A threshold of generations is defined to avoid ineffective explorations. In case no adversarial examples are found until the threshold is met, the input object is labeled as 'unmodifiable' and the process continues with the next file.

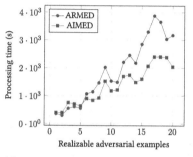
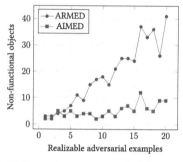

(a) Comparing average processing times.   (b) Comparing generation of valid objects.

Figure 5.3 Comparison between stochastic and GP approaches based on the processing time and generation of realizable objects. With AIMED, the processing time is reduced when searching for a larger number of adversarial examples (a), whereas the number of unrealizable examples is significantly decreased (b)

In comparison to the stochastic approaches explored in Chapter 4, genetic algorithms show improved convergence rates when searching for new adversarial members. As can be observed, AIMED works faster and more efficiently than ARMED when the expected number of adversarial examples scales. In fact, the gaps widens toward the end, at which the processing time of AIMED is at least twice as when searching 20 or more adversarial examples. Conversely, for five or fewer files, the processing times of both approaches are similar, with slightly better performance observed with ARMED, as shown in Fig. 5.3(a). This is likely because AIMED requires a few generations until it starts converging results in the same time that ARMED is able to randomly find a handful of evasive files.

Even better results are reported for the GP approach when generating functional adversarial examples. As shown in Fig. 5.3(b), to achieve 20 valid files, AIMED needs to generate nine additional corrupt files (29 in total), which accounts for a success rate of 68,9 %. ARMED, however, needs to generate 62 files (42 corrupt) to find 20 functional adversarial examples, accounting for a success rate of 32.25 %. Therefore, it produces four times more nonfunctional PE files than AIMED, which is a significant improvement in terms of plausibility.

By analyzing the results obtained with AIMED against a given target model, we speculate whether fresh attacks found during this process can evade further classifiers without extra computation. In other words, we investigate how the set of freshly generated adversarial malware examples would perform against unseen targets potentially evading the models.

5.4   Cross-evasion

After evaluating the performance of a GP-based approach against research models, we explored a more realistic scenario generating attacks against commercial classifiers, in which the adversaries have no knowledge of the target. The underlying idea is to assess whether attacks against the research model can be leveraged to identify the weaknesses of real-world classifiers. Furthermore, we investigated potential cross-evasions, in which the adversary generates only adversarial examples against one target and uses the same object to bypass multiple models. For our experiments, we created sets of adversarial examples and grouped them by target. Then, we used each set against different models to report cross-evasion rates. As targets, we chose along the research model the top commercial classifiers, namely, Kaspersky, Sophos, and ESET, with their default configurations and all updates available at the moment of the evaluation. Table 5.1 shows the results obtained.

Table 5.1 Cross-evasion rate comparison among selected research and commercial classifiers. Adversarial examples are grouped according to the initial target model and used to evaluate multiple classifiers without computing new attacks

	KASPERSKY	ESET	SOPHOS	RESEARCH
KASPERSKY	**100%**	18,39%	49,69%	25,47%
ESET	66,02%	**100%**	69,90%	38,83%
SOPHOS	41,74%	42,15%	**100%**	42,71%
RESEARCH	38,46%	48,07%	82,69%	**100%**

Our experiments demonstrate that cross-evasion is possible by finding successful adversarial examples against research models and later using them directly (without extra computations) to evade commercial classifiers, and vice versa. In total, 42.71 % of successful malware against Sophos effectively evades the research model. Moreover, the set of adversarial examples found against the research model shows promising evasion rates against its commercial counterparts, for instance, bypassing ESET and Sophos in more than 48 % and 82 % of the cases, respectively. In contrast, adversarial examples against ESET are less effective when targeting the research model (38.83 %). However, they display encouraging results against Kaspersky and Sophos, with more than 66 % evasion rates for each.

Overall, our results highlight that adversaries can use local and more accessible models to target commercial classifiers with relatively high evasion rates without having to query them individually to find weak spots. This means that adversaries may be able to compute attacks only once and leverage them, albeit with lower success rates, against further targets.

5.5   Summary

In summary, our experiments show that AIMED can converge twice as fast than stochastic approaches while duplicating the number of functional adversarial examples. This means that adversaries can leverage the optimization techniques to achieve successful attacks against static malware classifiers within minutes. Moreover, GP occasionally converges toward suboptimal solutions, which poses the risk of falling into local minima. In addition, genetic mutations need to be implemented carefully, since higher rates increase randomness and can compromise the evolution process.

Reinforcement Learning

6

Learning is a lifetime process, but there comes a time when we must stop adding and start updating.
—Robert Breault

After exploring evolutionary approaches, such as genetic algorithms (i.e., AIMED in Chapter 5), we observed that strong improvements are achieved compared to regular stochastic methods (i.e., ARMED in Chapter 4). In general, GP is significantly faster in identifying successful adversarial examples compared to stochastic approaches. However, it can often converge to local minima, in which less optimal solutions arise that are insufficient to generate valid adversarial examples. Therefore, further mechanisms need to be investigated.

The scope of the AML problem in the context of malware suggests that exploring techniques to identify optimal sequences of transformations seems to be in line with the goals of RL agents. To begin with, input objects and transformations are properly defined. Additionally, the fitness function (§ 5.2.2) consists of parameters that may be used to calculate a reward function.

However, earlier studies [Fan+19, Fan+20] indicate that RL agents tend to optimize toward stronger perturbations (i.e., UPX packing) with very large sequences, which may be easier to flag. Hence, our goal is to find efficient agents that can match the performance of related approaches with significantly shorter and more heterogeneous attacks.

We therefore introduce Automatic Intelligent Manipulations to Evade Detection with RL (AIMED-RL), a module that extends AIMED's capabilities, as shown in Fig. 1.1, by focusing on the automatic identification of novel adversarial examples using RL agents [LFR21].

© The Author(s), under exclusive license to Springer Fachmedien Wiesbaden GmbH, part of Springer Nature 2023
R. Labaca-Castro, *Machine Learning under Malware Attack*,
https://doi.org/10.1007/978-3-658-40442-0_6

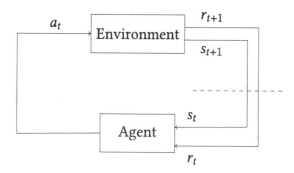

Figure 6.1 Representation of the MDP. In the malware context, the object reports its state s_t as a feature vector that contains essential information about the malware file. The actions a_t consist of a toolbox of byte-level transformations whereas the reward r_t is a function with multiple components including the classification rate

The idea of this module is to solve an optimization problem using RL agents. Thus, given an input state, a set of actions that maximize the reward function are determined. The process can be defined as a four-tuple Markov Decision Process (MDP) $(S, A, P_{s,a}, R_{s,a})$. Here, S stands for the *states space*, which is the feature space of the input object. A refers to a set of possible actions, also known as the *action space*, which are based on byte-level perturbations. $P(s, a)$ represents the *probability distribution* throughout the next states. Finally, $R(s, a)$ is the *reward function* obtained from transitioning states.

6.1   Methodology

The goal of the module is to train an agent for a number of turns (i.e., discrete timesteps) to interact with a dynamic environment. For every turn t, the agent choses an action $a_t \in A(s_t)$ and transitions from state s_t to s_{t+1} where $s_t \in S$. The agent thereby receives a reward $r_t = r(s_t, a_t)$ and updates policy π at step $t : \pi_t(s, a)$ (Fig. 6.1).

To learn the best action for each state, a Q-learning algorithm [WD92] can be implemented. The idea behind this model-free algorithm is to associate values with given actions on particular states to maximize the expected value of the reward over all states. Hence, the function adds the current reward with the maximum potential reward achievable:

$$Q(s, a) = r(s_t, a_t) + \gamma \max_{a_{t+1} \in A} Q(s_{t+1}, a_{t+1}) \tag{6.1}$$

where $\gamma \in [0, 1]$ represents the discount factor that adjusts the value of potential future rewards. The policy aims to prioritize *current* over future rewards. Hence, if γ is assigned a value close to zero, then this means that the agent will prioritize immediate rewards, whereas for higher γ values, future rewards will be more appreciated. Thus, it is important to find an optimal policy π that aims to maximize the expected return E starting from an initial state. In this context, we denote by $V^*(s_t)$ the maximum value that $V^\pi(s_t)$ can achieve in Eq. 6.2 as follows:

$$V^\pi(s_t) = E\left[Q(s_t, a_t | s_t)\right] \tag{6.2}$$

Note that having a matrix to store the values of state-action pairs becomes quickly complex to maintain since the values grow exponentially. Hence, a neural network can be implemented to predict the Q-values using the agent's feedback.

We now define each element of the RL scenario in the context of malware and explain how states and actions interact in the system. Let us also define the reward function based on domain parameters.

6.1.1 State

A given state s consists of the features extracted from the input object. For every change in the environment, namely, actions taken within a state, the next s_t is defined on the basis of the features extracted from the input object at every iteration.

6.1.2 Action

Each action a maps a byte-level transformation, as introduced in § 3.3.3. In this scenario, the agent can sample from the defined set and change the state of the input object by injecting transformations into the adversarial example.

6.1.3 Reward

Inspired by AIMED's fitness score in Chapter 5, we introduce reward as a linear function of three parameters: detection, distance, and similarity. The functionality parameter is left out in AIMED-RL, since the integrity of the adversarial examples is verified at the end of the pipeline for efficiency reasons. A maximum reward value is set for each parameter so that $r_{max} = 10$.

Detection

The parameter r_{det} varies depending on the label returned by the classifier. For $r_{det} = 0$, the object is labeled as malicious, whereas for $r_{det} = 10$, the object is labeled as benign.

Distance

The parameter r_{dis} is based on the number of turns multiplied by a factor t, which incentivizes the agent to aim for larger rewards when the sequence of transformations is larger (up to $t_{max} = 5$). The limit is based on the optimal trade-off between the number of transformations and evasiveness while preserving the integrity of the object (§ 3.3.4):

$$r_{dis} = \frac{r_{max}}{t_{max}} * t \qquad (6.3)$$

Similarity

The parameter r_{sim} consists of a byte-level comparison between two objects: the file provided as input and the adversarial example generated after an action is taken. The larger the difference between these two objects, the more diverse the attacks expected.

$$r_{sim} = (1 - s_{ratio}) * r_{max} \qquad (6.4)$$

Reward function

To adjust the importance among the different parameters while tuning the model for better performance, we introduce ω. Overall, the linear combination in Eq. 6.5 summarizes the reward in the context of malware:

$$R = R_{det} * \omega_{det} + R_{sim} * \omega_{sim} + R_{dis} * \omega_{dis} \qquad (6.5)$$

We therefore introduce two different strategies, *standard* and *incremental*, based on empirically defined weight distributions. The former allocates the same weight to each reward, whereas the latter shifts most of the attention toward detection.

Table 6.1 Definitions of strategies according to different weight distributions for the reward function. The standard strategy allocates the same weight throughout all components of the reward function, whereas the incremental strategy assigns higher preference to rewards depending on the detection rates

STRATEGY	R_{DET}	R_{DIST}	R_{SIM}
Standard	0.33	0.33	0.33
Incremental	0.50	0.30	0.20

6.1.4 Penalty

Occasionally, the agent tends to enter repetition loops in an attempt to find an optimal solution, given that iterating over the same actions may maximize the reward function [And+18]. From the malware perspective, however, adding the same transformation iteratively can generate weaker attacks, since such attacks can potentially be flagged by the classifier. Hence, in this section, we explore a penalization technique to encourage the agent to find more heterogeneous sequences.

$$R = \begin{cases} R & \text{for } \rho = 0 \\ R * 0.8 & \text{for } \rho = 1 \\ R * 0.6 & \text{for } \rho > 1 \end{cases} \quad (6.6)$$

where ρ is the penalty factor that increases when duplications are registered, which explicitly aims to discourage repeated actions and, hence, minimize the impact of stronger transformations.

6.1.5 Model Parameters

Overall, initial approaches implementing RL agents in the context of malware have explored the policy-based approach Actor Critic with Experience Replay (ACER) [And+18]. Recent attempts have also shown preference for value-based networks as promising strategies to generate adversarial malware examples [Fan+19, Fan+20]. Therefore, in this work, we choose a Deep Q-Network (DQN) algorithm using noisy nets as an exploration strategy with selected enhancements [Mat+18].

Specifically, we implement a Distributional DQN (DiDQN) [BDM17] and combine it with a Double DQN (DDQN). This is because the regular DQN algorithm has been shown to overestimate action values in certain cases [VGS16]. The DiDQN consists of two hidden layers and 64 nodes with parameters $V_{min} = -10$, $V_{max} = 10$, $N_{atoms} = 51$, and it aims to learn the distribution of rewards instead of the expected value. The DDQN is configured with $\alpha = 0.001$, $\beta_1 = 0.9$, $\beta_2 = 0.999$, $\epsilon = 0.01$, and Adam [KB14] as an optimizer. A detailed comparison of our agent, a Distributional Double DQN (DiDDQN), and current approaches highlighting optimizer, discount factor, and exploration algorithms is presented in Table 6.2.

Table 6.2 Overview of RL implementations in the literature. While most approaches use similar discount factors, agents and explorations differ across different studies in the context of malware. Agents based on DQN variations are the most effective in maximizing the evasion rates

APPROACH	AGENT	OPTIMIZER	DISCOUNT	EXPLORATION
Fang et al., 2019	DDQN	Adam	0.99	$\epsilon - greedy$
F., Zeng et al., 2020	DuDDQN	RMSProp	N/A	Boltzmann
Anderson et al., 2018	ACER	Adam	0.95	Boltzmann
AIMED-RL	**DiDDQN**	**Adam**	**0.95**	**Noisy Nets**

6.2 Evaluation

For the training step of our experiments, we sampled 4,187 PE files from the FAME dataset, as defined in § 3.3.2. For the testing step, we defined a holdout subset of 200 files.

6.2.1 Reset

As outlined in § 6.1, the agent is limited to select up to five actions per file. However, before moving on to the next episode, we introduce a reset mechanism, which aims to support the idea that the order of the sequence plays a more important role than its length. Once the maximum length of the sequence is achieved, the input object returns to its original state and a new attempt starts within the same episode. Therefore, the agent has a second opportunity to identify a successful adversarial example in case the first fails.

6.2.2 Diversity

Earlier studies indicate strong domination of *upx_pack* that negatively impacts the results obtained by the agent [Dem+20, Fan+20]. Hence, we aim to improve the diversity of the sequences in order to discourage the agent from repeating the same actions and producing homogeneous attacks.

Overall, implementing the *reset* mechanism combined with a *penalty* technique allows for more diverse sequences of transformations. While *upx_pack* still dominates the spectrum as expected, new actions, such as *imports_append*, *section_append*, and *upx_unpack*, acquire higher visibility, as shown in Fig. 6.2.

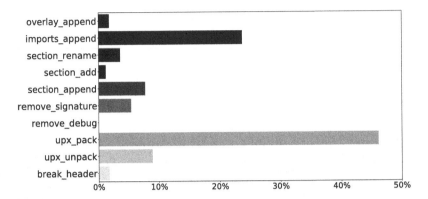

Figure 6.2 Distribution of transformation usage based on best performing agent. We observe that after introducing the penalty technique, the dominance of upx_pack transformation is still over than 40%, yet further actions such as $imports\_append$ and $section\_append$ are often chosen by the agent and therefore deemed essential to generate effective adversarial examples

We nonetheless expect certain perturbations to be dominant depending on how the features of static malware models are extracted. *Packing* or *compressing* transformations are good candidates to be strong, since the feature domain is essentially determined by the structure of the PE file and such techniques strongly affect the structure. Thus, we advocate for decreasing the dependency of the agent on single transformations and thereby encouraging diversity.

However, it is important to address not only homogeneous sequences but also attacks [Dem+21, HT17] that heavily rely on actions that can potentially be flagged with preanalysis techniques (i.e., appending bytes at the ends of sections). In these cases, such transformations may be removed from the input object before the prediction takes place.

6.2.3 Evasion Rate

For each testbed we trained 10 agents during 1,000 and 1,500 episodes[1] respectively, using both the *incremental* and *standard* strategies, as defined in Table 6.1. Further-

---

[1] Note that we have empirically defined the upper range of the episodes for training RL agents, since we have not detected any improvements after 1,500 episodes. Likewise, the agents required on average at least 1,000 episodes to yield optimal results.

Table 6.3 Comparison of evasion rates among agents with two weight strategies, *incremental* and *standard*, combined with a penalty (WP) or no penalty (NP) technique to discourage homogeneous attack sequences

STRATEGY	EPISODES	AVG. EVASION	BEST AGENT
Incremental (WP)	1000	23.52%	40.00%
	1500	20.35%	33.84%
Incremental (NP)	1000	18.78%	30.81%
	1500	23.06%	35.35%
Standard (WP)	1000	21.07%	35.86%
	1500	26.29%	43.15%
Standard (NP)	1000	21.6%	30.0%
	1500	23.74%	41.41%
Random Agent	—	21.21%	24.62%

more, we trained agents with a penalty (WP) and with no penalty (NP) technique. We also include an additional random RL agent as baseline for comparison. The results are presented in Table 6.3.

Agents trained with both reward distribution strategies generally achieve better average evasion rates than random agents. Moreover, the best agent typically outperforms the random policy regardless of the settings used. When comparing between WP and NP approaches, adding a penalty provides a clear advantage and leads to the highest performing agents during the experiments.

While average evasion yields better results for NP with 1,500 episodes and 23.06 % evasion over 20.35 % using the penalty technique, the agent trained with WP needs 1,000 episodes to achieve the best results using the *incremental* strategy, namely, 23.52 % average and 40 % evasion for the best agent.

Similarly, agents using the *standard* weight distribution strategy score on average better with WP than with NP, although they require 500 episodes more to achieve their best performance. In fact, the best scoring agent reported 43.15 % evasion on the holdout set using 1,500 episodes and the *standard* strategy.

After integrity verification, we determined that the quality of the sequences of transformation is very high. In other words, 97.64 % of the adversarial malware examples generated are valid. Thus, by adjusting the evasion rate according to functionality, the best agent reported 42.13 % realizable evasions.

Therefore, AIMED-RL yields high evasion results while working with an efficient approach using a limited number of episodes and a very short sequence of transformations. Our agent needed to be updated nearly 13,900 times, whereas similar approaches [And+18] reported up to 50,000 modifications and lower evasion rates.

Table 6.4 Evasion results of AIMED-RL compared against different approaches. The Integrity Verification Stage (IVS) indicates whether the integrity of the file is verified after the adversarial malware example is generated

APPROACH	SPACE	REWARD	PERTS.	MODEL	IVS	EVASION
Y. Fang et al., 2019	4	R_{det}, R_{dist}	80	LGBM	✓	46.56%
Z. Fang et al., 2020	218	R_{det}, R_{dist}	100	DDNet	✓	19.13%
Anderson et al., 2018	11	R_{det}	10	LGBM	✗	16.25%
AIMED-RL	**10**	**$R_{det}, R_{dist}, R_{sim}$**	**5**	**LGBM**	**✓**	**42.13%**

Table 6.4 presents the summary of a comparison between the results of AIMED-RL with similar studies. Notably, the RL approach is frequently implemented in the literature to classify PE binaries. Although Z. Fang et al. [Fan+20] trained a new model, they reported comparable performance to that obtained by Anderson et al., whose work is also evaluated.

Among all approaches, Y. Fang et al. [Fan+19] reported the highest evasion rate of 46.56%, returning valid adversarial examples that have been injected with up to 80 perturbations. However, limited implementation details are provided to reproduce the attack and perform a deeper comparison.

Z. Fang et al. [Fan+20] released the dataset without the algorithms. Thus, to provide a better comparison, we evaluated our best agent with their released dataset. The agent achieved 16.07% evasion. As opposed to 100 transformations proposed in their approach, our agent injected only five. The integrity verification stage also revealed 84% valid adversarial examples. The authors also indicated the use of IDA Pro [Hex21] as an additional verification step. While this approach can help guarantee valid adversarial examples, it may involve manual validation and, thus, can significantly raise the cost of finding effective adversarial examples.

Therefore, AIMED-RL focuses on fewer transformations as the cost to find adversarial examples injecting 15 to 20 times larger sequences may be arguably more expensive and significantly raises the chances of producing nonfunctional malware objects.

6.3   Summary

In summary, AIMED-RL reports high performance results when generating adversarial malware examples. Our experiments show that the agent requires 1,500 episodes (3,000 if *reset* is considered) and only five transformations to automatically generate strong attacks against malware classifiers. In addition, introducing penalty techniques considerably reduces the dependency on unique transformations, which leads to improvements in the diversity of sequences and more effective attacks. Therefore, RL agents seem to be an effective alternative to previously discussed GP and stochastic strategies in terms of input-specific attacks. Furthermore, AIMED-RL outperforms similar state-of-the-art approaches that rely on RL by significantly reducing the attack sequence and increasing the evasion rate while producing realizable adversarial examples.

Universal Attacks

7

Science provides an understanding of a universal experience. Arts provide a universal understanding of a personal experience.

—Mae Jemison

In Chapters 4—6, we explored input-specific attacks, in which a committed adversary needs to calculate the best sequence of perturbations for every object. In other words, when the input file changes, a fresh attack is computed in an attempt to successfully bypass the classifier. Although this is highly effective, adversaries can now take the game to the next level. For example, what if one single attack can induce the model to return a wrong classification for an important fraction of the input objects? This class of adversarial attacks is known as UAPs, whose existence have been highlighted by Moosavi-Dezfooli et al. [Moo+17]. UAPs systematically expose the weaknesses of models and represent significant risks since realistic attacks are conducted across different domains, including computer vision problems [Ath+18, Bro+17, Co+19, KO18, Mop+18] audio applications [Nee+19, Abd+19], and ad-blocking [Tra+19]. A recent study [Hou+20] has also explored such attacks in the Android malware domain, but with limited application to the feature space only.

Therefore, we introduce the Generate Adversarial Malware Examples with Universal Perturbations (GAME-UP) module and present it as the last attack of the problem space in Fig. 1.1. The goal of this module is to maximize the impact of a single sequence of transformations across a large portion of input objects producing adversarial examples without the need to compute individual attacks [Lab+22]. Thus, it allows identifying systemic vulnerabilities within ML-based malware classifiers.

R. Labaca-Castro, *Machine Learning under Malware Attack*, https://doi.org/10.1007/978-3-658-40442-0_7

7.1   Requirements

In this section, we extend the assumptions in the previous module (§ 4.1) by including two additional requirements that are relevant for GAME-UP, since the process of finding universal attacks differs from input-specific approaches.

- *Datasets*: The search for UAPs is divided into two steps, which each require separate datasets. In the *exploration* stage, universal perturbations are searched and identified, whereas in the *validation* stage, the selected UAP is tested using a holdout dataset to ensure that the evasion rates are consistent.
- *Confidence score*: The implemented classifier returns the soft labels consisting of confidence ratios instead of prediction classes. This information is relevant for the UAP search, because multiple sequences are evaluated and hard labels (i.e., detected/not detected) provide only a binary response regarding the success of the attack, whereas soft labels provide comparable values indicating proximity to the decision boundary.

7.2   Universal Evasion Rate

To assess our results, we define the *effectiveness* of UAP attacks with regard to the Universal Evasion Rate (UER) determined by a set of inputs $\mathcal{X}$:

$$\text{UER} = \frac{|\{x \in \mathcal{X} : \arg\max g(x + \delta) \neq y \in \mathcal{Y}\}|}{|\mathcal{X}|} \tag{7.1}$$

The UER calculates for which inputs in $\mathcal{X}$ the model returns errors when the universal perturbation δ is employed.

7.3   UAP Search

The goal of GAME-UP is to explore, identify, and validate an optimal sequence of transformations, known as UAPs, which maximize the UER (Eq. 7.1) over a set of input objects.

We start the process by applying each of the 10 transformations (§ 3.3.3) to every input object $x \in \mathcal{X}$ in the *exploration set*. Next, we calculate for each transformation a minimum confidence rate[1] and select the perturbation that minimizes the score as the first transformation in the sequence. The process is repeated until the expected length of the universal adversarial sequence is reached. The search rounds r are determined by a product of the magnitude of the *exploration set* E, sequence of transformations $\mathbf{T}$, and set of perturbations available $\mathcal{T}$—i.e., that is, $r = |E| \cdot |\mathbf{T}| \cdot |\mathcal{T}|$. Then, the sequence of perturbations that maximizes the UER is selected as the UAP. Finally, the UAP's effectiveness is verified throughout the *validation set* to compare the evasion rates across a holdout dataset, after which the performance is reported.

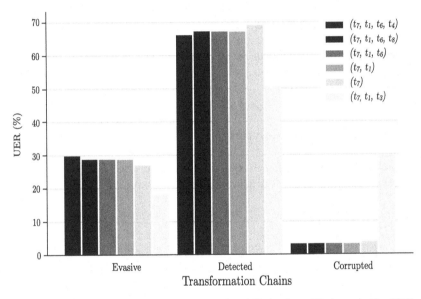

Figure 7.1 Evaluation of the best candidates for UAP in the validation set. The UAP, (t_7, t_1, t_6, t_4), achieves 298 adversarial examples, whereas the UAP (t_7, t_1, t_3) shows how the order of the sequence significantly impacts the success of the sequence

---

[1] If the target model does not grant access to soft labels, then a set of commercial classifiers are used to provide the confidence rates when classifying binaries [Vir22].

7.4   Evaluation

In this section, we validate the effectiveness of the most successful UAP candidates. We run our experiments with $|\mathbf{T}|$ of 1, 4, and 10 but evaluate UAPs of length 4, since we have not registered improvements with longer sequences and minimal attacks are desired to ensure *plausibility*.

During the exploration process, three sequences reported the highest UER, whereas two sequences yielded similar evasion rates with shorter sequences. A low scoring candidate is also chosen as a control to study less evasive sequences.

From the six candidate sequences, (t_7, t_1, t_6, t_4) is selected as the UAP, reporting 298 valid adversarial examples from 992 files[2], with a score of 30%, as shown in Fig. 7.1. The following UAP candidate (t_7, t_1, t_6, t_8) returns 288 adversarial malware examples, reporting an evasion score of 29%. The third most successful sequence (t_7, t_1, t_6), reports the same UER, although it requires only three transformations. As can be observed, since the three highest performers share the same prefix (t_7, t_1, t_6), the last perturbation seems to have little influence on the overall performance. The fourth candidate (t_7, t_1) reports 287 adversarial files, which corresponds to 29% UER, whereas the single transformation t_7 returns 270 adversarial examples (27.2% UER). Here, t_7 corresponds to the byte-level perturbation *upx_pack*, which is visibly strong and dominates the attacks across the modules of FAME (cf. Chapter 6), given that the LightGBM model and static malware classifiers in general focus on the structure of the PE file. Additionally, the distribution of header information for malicious files packed with UPX is consistent with being benign irrespective of maliciousness. Besides, unit vectors consisting of a packing transformation can be potentially simpler to flag. Thus, sequence t_7 reports the highest detection rate among all UAP candidates, which indicates that diversity is relevant for the success of adversarial examples.

The remaining sequence (t_7, t_1, t_3) is used to show how potentially successful prefixes, that is, (t_7, t_1), do not necessarily lead to effective adversarial examples. Appending transformation t_3 (adding a new section to the PE) can increase the number of corrupt files drastically, hence reducing the UER by almost 40%, whereas t_6 (removing debug information) contributes to the UER, making the UAP candidate the third strongest. Such behavior is eventually explained by the nature of the change in the binary. Adding a new section to the file has a more aggressive implication for the structure of the binary than manipulating debug information from certificate tables.

---

[2] Note that the size of the *exploration set* is 1,000 PE files, however eight were excluded because to parsing errors or positive detection by the baseline model.

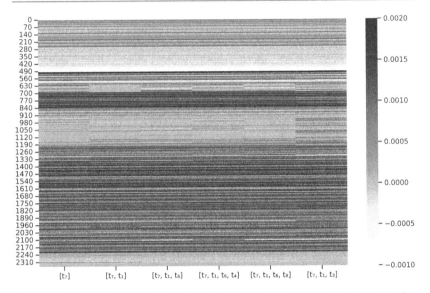

Figure 7.2 Average of the delta variation for adversarial examples in the *exploration set* after modification with each UAP candidate. The y-axis depicts the mapping of the feature space. While the distribution of features is similar across candidates, the control sequence (t_7, t_1, t_3) shows a clearly different pattern in the feature domain

We finally explore the *delta variation* in the feature-space for every UAP candidate presented and report the results in Fig. 7.2. From left to right, the figure shows the best five UAP candidates, with the last one being the least effective sequence. Candidates with the highest evasion rates show similar distributions across their features, supporting the classifier's evasion. However, the sequence (t_7, t_1, t_3) clearly shows a different pattern, especially between features 800 and 1,000. This range of features corresponds to information about sections (e.g., names and sizes) that are presumably affected when new sections are created.

7.5 Summary

Overall, GAME-UP shows that finding valid universal examples using greedy algorithms in the context of PE malware is possible. Unlike input-specific attacks, UAPs need to be computed only once and can be used across a large set of objects, hence substantially reducing the complexity of producing new attacks. Therefore, adversaries can leverage novel mechanisms to reduce the effort required to evade malware classifiers.

Similarly, UAP attacks are useful in systematically assessing the weaknesses of models. A single sequence of transformations can reveal relevant information about the robustness of the predictions without requiring more expensive individual processing until a number of adversarial examples are identified.

Part IV

Feature-Space Attacks

Gradient Optimization

8

Progressing at a snail's pace is still progress, and slow progress is better than no progress. Never be stagnant.

—Richelle E. Goodrich

Since further domains, such as computer vision, highly consider white-box settings using, for example, gradient attacks, we also investigated these for malware classification. In this section, we explore how to generate in the feature space adversarial examples that can evade malware classifier models with gradient attacks, using a CNN to detect Windows PE files [Pra18].

First, let us present Gradient Relevant Injection to Portable Executables (GRIPE), in which we leverage the toolbox from ARMED (§ 4.2) to generate adversarial examples using gradient information [LBD19], which appears as the first feature-space attack in Fig. 1.1. Instead of using random sequences of transformations, we maximize the evasion rate of every perturbation by evaluating the gradient value extracted. This process allows generating optimal adversarial examples faster, since the adversary would have access to the network and its weights.

8.1   Convolutional Neural Networks

CNNs are inspired from the connectivity pattern of the human brain and the organization of the visual cortex, where a set of neurons individually respond to stimuli in certain regions of the visual field that later overlap to cover the entire spectrum. Instead of flattening the input object (e.g., an image) to a column vector, a CNN performs a series of steps to extract the spatial and temporal dependencies of the object. Hence, CNNs are a special type of feed-forward neural networks whose architecture

R. Labaca-Castro, *Machine Learning under Malware Attack*,
https://doi.org/10.1007/978-3-658-40442-0_8

consists of three main types of layers: (i) a convolutional layer, which is responsible for extracting the basic elements of the image until high-level features are obtained; (ii) a pooling layer, which reduces the spatial size and extracts dominant features from the object; and (iii) a fully connected layer, which learns the nonlinear combinations of high-level features, leading to the classification of the object. In this context, the domain may be extended from image and video processing to multiple contexts, including malware classification [Tob+16].

8.2   Methodology

In this work, a CNN was trained using the $EMBER$ dataset (§ 3.3.2), yielding an F-score of .94 at 5 % FPR. Fig. 8.1 shows the AUC-ROC. Overall, the model can potentially be further tuned to improve the performance and stabilize the loss function throughout training.

Generally, the toolbox of the ARMED module (§ 4.2) is responsible for generating the adversarial malware examples. The workflow consists of an input object x receiving a sequence of transformations to generate x'. The generated example is then assigned a confidence score. If $g(x') < g(x)$, then the process is deemed successful and an adversarial example is generated.

Instead of generating adversarial examples using a black-box approach (i.e., attacking the LightGBM model), we target a CNN under white-box settings. In this attack scenario, the adversaries' capabilities are increased, since they have full knowledge of the model and its hyperparameters (§ 3.2.2). Hence, for every transformation injected into the input object, a gradient value can be extracted.

Following the conventions adopted in § 3.1, we define the notation as $g : \mathcal{X} \longmapsto \mathcal{Y}$ in the feature space $x \in \mathcal{X}$. Every adversarial malware example generated receives a label output by a continuous discriminant function $f : \mathcal{X} \in \mathbb{R}$. Thus, we assume that $g(x) = 1$ for $f(x) > 0$ and $g(x) = 0$ if $f(x) < 0$. Attacks are bounded by g_{max} since perturbations are constrained by the maximum distance: $g : \mathcal{X} \times \mathcal{X} \longmapsto \mathbb{R}^+$. The goal here is to minimize $f(\cdot)$, hence the estimation $\hat{f}(\cdot)$, to obtain an adversarial example x' based on the following equation:

$$e = \arg \min_{s} \hat{f}(\cdot) \ \ s.t. \ \ g(x, x') \leq g_{max} \quad \quad (8.1)$$

We approach this problem using the gradient descent technique, which aims to minimize the error (loss) until achieving the best set of parameters for the model. While such algorithms report favorable performance, local optimization may fail given the nature of the discriminant function. In our case, we attempt to minimize the risk of facing such issues by training the model with a rather large dataset.

We now introduce our solution, whose goal is to generate successful adversarial examples in the feature space based on gradient information extracted from the target model. When generating malware examples, since PE files can experience corruption after subtracting parts from their structure, we focus only on adding or modifying instead of removing information.

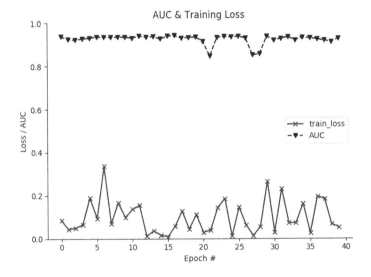

Figure 8.1 AUC-ROC and training loss reported by the CNN. The model outputs a high performance, whereas the train loss may be further stabilized

The process of producing adversarial examples consists of obtaining a score for each transformation and calculating the gradient for the following iteration. Having access to the neural network makes the identification of the optimal sequence of transformations significantly more efficient. In terms of integrity, the adversarial malware examples produced are valid, since transformations are transferred from real-world attacks into the feature space.

Unlike in LK scenarios in which black-box classifiers are generally targeted, PK attacks provide a better understanding of the intrinsic behavior of the classifiers, which especially allows measuring the impact of real-world transformations. However, such attacks do not produce problem-space objects, and malware examples are represented in the feature space instead. The idea in this case is to leverage the gradient information to explore the weaknesses of the target model to adversarial malware examples originating in the problem space. The knowledge retrieved from this process can potentially be used to improve the resilience of malware classifiers toward more robust models against adversarial attacks. Such attacks can be leveraged to evaluate cross-evasion capabilities across multiple model architectures. Note that, in some cases, the adversary may not be able to access the internal parameters of a model but can train a surrogate instead. By querying a target model that provides only hard labels, a new classifier can be trained, which can then be used to explore gradient-based attacks.

8.3   Summary

In this section, we proposed GRIPE, a gradient-based attack against CNNs trained with Windows PE files. Our approach leverages the information extracted from PK scenarios into building optimal sequences of transformations that can generate effective adversarial attacks. The resulting adversarial examples are produced by transferring problem-space perturbations into the embedded domain and, thus, generating objects as valid representations in the feature space. Unlike most approaches in the context of malware, we leverage the information of real byte-level perturbations in the continuous space instead of focusing on activating features (i.e., flipping bits from zero to one) throughout the binary representations of malware objects. As a potential improvement, the efficiency of the process can be further enhanced by implementing an additional neural network to estimate the distribution of the features rather than extracting and computing the gradient value for every input object throughout the process.

Generative Adversarial Nets

Individually we are one drop; but together we are an ocean.
—Ryunosoke Satoro

Goodfellow et al. [Goo+14] introduced the concept of GANs, which consist of two neural networks trained by competing (collaborating) with each other in order to optimize opposite goals in a zero-sum game framework. Although in this scenario one network profits from the other's loss, GANs become better at their predictions by means of cooperation rather than competitiveness. The generator learns the statistics from the training set to produce new synthetic data, whereas the discriminator assumes the role of an evaluator that determines whether the input received from the former is real or synthetic.

Hence, the minimax game from both *generator* G and *discriminator* D can be defined with value function $V(G, D)$ as follows:

$$\min_{G} \max_{D} V(G, D) = \mathbb{E}_{\mathbf{x} \sim p_{data}(\mathbf{x})}[\log D(\mathbf{x})] + \mathbb{E}_{\mathbf{z} \sim p_{\mathbf{z}}(\mathbf{z})}[\log(1 - D(G(\mathbf{z})))] \quad (9.1)$$

where G receives an input vector z and the D is trained with two input data batches, namely, x sampled from real probability distribution p_{data} and data generated from $G(z)$.

R. Labaca-Castro, *Machine Learning under Malware Attack*,
https://doi.org/10.1007/978-3-658-40442-0_9

9.1   Gans in Security

Following their introduction in 2014, GANs received substantial performance improvements with the generation of hyperrealistic objects, animals, and human faces [KLA19; Lar+21]. Additional approaches have also been explored beyond image and video processing, including the security domain. Anderon et al. [AWF16] proposed DeepDGA, an approach to evade a web domain generation classifier that can detect when domains are generated either by humans or automatically via domain generation algorithms.

Rigaki et al. [RG18] introduced an approach to bypass IPSs. They mimicked traffic from the Facebook messenger application by manipulating a malware code, which suggests the usability of GANs in the network traffic domain and their effectiveness in reducing the detection rates.

Specifically in the context of malware, Hu and Ying [HT17] presented Mal-GAN by simulating a malicious GAN. Their goal was to generate malicious software representations in the feature space that successfully evade classification. To achieve evasion with a more realistic scenario, they implemented a surrogate model to fit the real black-box malware classifier, which in turn acted as a *discriminator*. Then, they leveraged the gradients from the freshly trained model by the *generator* to produce adversarial examples. In this scenario, the authors reported perfect evasion. However, it requires substantial knowledge from the adversary (i.e., being fully familiar with the feature space), which makes the attack less realistic despite targeting black-box classifiers. Additionally, the representation of PE files is rather simplistic in vectors of short length consisting of API calls only, dismissing critical information from the binaries. Attacks in binary feature spaces are presumably more successful than continuous ones in generating effective adversarial examples, since flipping bits requires little computation compared to processing more complex representations of the PE file.

We therefore propose Generative Adversarial Intelligent Network to Evade Detection (GAINED), which is the next attack in the feature space in Fig. 1.1. GAINED consists of a GAN approach that extends the work of Hu and Ying to compute more complex feature representations of PEs based on the toolbox (§ 3.3.3) with byte-level transformations transferred from the problem space.

9.2   Methodology

In this module, we implemented a GAN to generate adversarial examples that combine real transformations with malware representations to evade detection, as shown

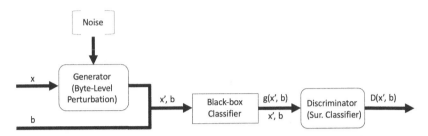

Figure 9.1 GAN-based workflow for malware evasion approach. Unlike regular generative networks, in this context the discriminator is implemented as a surrogate classifier, which fits the black box model (target) that is in charge of providing labels to the objects

in Fig. 9.1. The representation of malware examples is stored into a vector that includes a substantial representation of the binary, including header's metadata, file strings, and import and export tables.

The *generator* receives a malicious input object x along with *noise* that represents a vector of transformations. Then, an object x' returns as an adversarial example and together with a benign object b is sent to the black-box malware classifier, which detects both files $g(x', b)$. The objects and the assigned labels are then used as input by the *discriminator*, which acts as a surrogate classifier and outputs the probability that the inputs are malicious as $D(x', b)$. Under this settings, both the *generator* and the *discriminator* are feed-forward neural networks, whereas the malware classifier is a black-box model. The *discriminator* needs to fit the malware classifier to acquire a similar behavior. Hence, once an adversarial example is produced for the *discriminator*, it can also evade the black-box model.

As opposed to traditional GANs, the *generator* and *discriminator* are trained to improve the ability of the former to produce an adversarial malware example that evades the latter in order to bypass the black-box classifier. Therefore, the attacks are limited by the ability of the *discriminator* to properly fit the malware classifier. We also bound the *generator* to produce sequences of perturbations of length n that map 10 transformations for each case. Hence, the selected adversarial sequences reach combinations up to 10^n, which provides the *generator* with flexibility to extend the spectrum to find effective attacks.

Our approach leverages the ability of GANs to generate adversarial examples using Windows PE files. For this approach to work effectively, malicious and benign software files are required to train the GAN. Overall, we broadly explore the following three steps.

First, we define feed-forward neural networks according to the problem settings. Since minimal changes in the environment can affect the networks, it is important to tune the parameters until stable training is achieved. Moreover, while increasing the number of nodes can better adjust the input, it can also rapidly overfit the data. In terms of structure, it may be useful to add new layers until no improvements are observed in the test error [Ben12]. Second, we use the network that acts as a surrogate, *discriminator*, to fit a black-box GBDT (§ 3.3.1) in order to update its weights and leverage gradient descent. Similarly, in the third step, the feed-forward network, *generator*, updates its own weights depending on the gradient input from the surrogate classifier.

9.3   Summary

Overall, GANs are highly efficient in generating powerful adversarial examples across multiple domains, hence laying the foundation for a variety of new applications. Adversaries can exploit such advances and improve their own capability to generate adversarial examples in the context of malware.

In this section, we presented GAINED, a GAN-based approach that explores the strength of GANs combined with malware generation for Windows PE files. While more complex malware representations do not achieve perfect evasion rates compared to simpler alternatives observed in the literature, this paves the way for leveraging the use of GANs to extend the attack capabilities of adaptive adversaries in the context of PE malware. As a next step, to achieve more realistic attacks, a combination between binary and continuous features may be useful, allowing the generator to produce novel attacks in one section while protecting the feasibility of the object in the other.

Part V
Benchmark & Defenses

Comparison of Strategies

10

Don't compare your beginnings to someone else's middle.
—Tim Hiller

Having properly introduced each module of FAME, we now proceed to evaluate the problem-space attack suite with a holistic approach in order to better understand the advantages and disadvantages of each strategy. As depicted in Fig. 1.1, our goal is to create an initial benchmark to evaluate adversarial examples in the context of malware using PE files. Notably, since gradient-based approaches (i.e., GRIPE in Chapter 8) and GAN attacks (i.e., GAINED in Chapter 9) generate feature-space adversarial examples, we did not include them in the main suite of realizable attacks implemented in FAME. Generally, feature representations of malware objects produced by these approaches tolerate stronger modifications and larger sequences, given that they do not undergo integrity verification stages. Therefore, they are not comparable to the limitations of byte-level perturbations in real software binaries.

10.1 Benchmark Settings

From the analysis of the results of the experimental evaluation, it can be observed that the identification of a better approach for each case depends on defined settings and specific scenarios. For example, stochastic approaches are a good option to identify weaknesses in the target model if only a handful of adversarial examples are expected. However, scaling adaptive attacks requires learning from previous mistakes to be efficient (e.g., GP- and RL-based approaches), which is a limitation for such strategies. Hence, we carefully evaluated which attack performs best in each scenario.

© The Author(s), under exclusive license to Springer Fachmedien Wiesbaden GmbH, part of Springer Nature 2023
R. Labaca-Castro, *Machine Learning under Malware Attack*,
https://doi.org/10.1007/978-3-658-40442-0_10

For the following comparison, we evaluate all attacks using LightGBM as target model, defined in § 3.3.1, and sample input objects from the FAME dataset (§ 3.3.2). For uniformity reasons, we set an upper boundary of 100 adversarial examples so that each approach generates the same number of PE files. Following the best trade-off between reaching high evasion rates and preserving the integrity (§ 4.3), we specify $|\mathbf{T}| = 5$. We hereby report the results in Table 10.1.

While stochastic approaches, such as ARMED (Chapter 4), offer fast convergence for a limited number of adversarial examples, they report the lowest evasion rates among all modules, as observed in the comparison table. AIMED (Chapter 5), on the other hand, implements GP and achieves a fourfold improvement in terms of evasion while preserving 30% more integrity of the input objects. This significant achievement in terms of performance is associated with slightly swifter processing times, rendering AIMED the fastest module.

Table 10.1 Comparison of FAME modules. The time variable is the number of hours needed on average to generate and evaluate 100 adversarial malware examples

MODULE	TECHNIQUE	TRAINING	TIME	VALID	FILES	EVASION
ARMED	Stochastic	No	0.87	61.0%	100	10.1%
GAME-UP	UAP	Heur.	9.57	99.6%	5,000	33.6%
AIMED	GP	Heur.	0.82	91.0%	100	42.0%
AIMED-RL	RL	Yes	1.37	97.6%	13,908	42.1%

In terms of universal attacks, Generate Adversarial Malware Examples with Universal Perturbations (GAME-UP) provides a favorable overview based on the effectiveness of single-time computed attacks. In other words, calculating the ratio of adversarial examples among sequences of length five achieves an evasion score of 33.6%. For comparison reasons, aiming for similar experiments with the rest of the modules, the evasion rate of GAME-UP is based on the number of adversarial examples with respect to $|\mathbf{T}| = 5$. However, if the reported result pertains to all 5,000 PE files generated, then the evasion rate drops to 29.5%. Interestingly, even after the evasion scores are adjusted, Universal Adversarial Perturbation (UAP) attacks still pose a significant threat against malware classifiers, proving that they are a highly efficient strategy in finding optimal sequences of transformations. Moreover, the attack reports almost perfect integrity with only 0.4% of nonfunctional files. However, finding the optimal UAP candidate is an expensive process that requires more time, at least seven times longer, than that required to compute the longest input-specific attack.

Overall, the GP and RL approaches are the best strategies that can be used when searching for a set of adversarial examples without substantially compromising the integrity of the files. Although the performance levels of both attacks are very close, the latter reports an improvement of 6.6% in generating valid adversarial examples compared to the former. At the same time, GP requires no training, and thus it saves 40% of the time at almost no evasion cost (0.01%). On the other hand, RL requires the generation of more than 13,000 files to be ready after the training stage. Moreover, learning agents include a large number of hyperparameters to tune until finding the best settings, whereas evolutionary algorithms may be simpler to implement and start producing results.

10.2 Summary

Overall, every approach shows certain strengths and weaknesses, in which finding the best module largely depends on the use case and constraints defined. For example, if a model needs to be quickly assessed against evasion, then a single adversarial example would suffice, and hence ARMED should be a reasonable choice. However, to further evaluate vulnerabilities, a larger set of files need to be generated to assess different aspects of the classifier. Thus, an approach that learns over time is preferred in this scenario, such as AIMED or Automatic Intelligent Manipulations to Evade Detection with RL (AIMED-RL). The choice depends on the trade-off between performance and functionality, namely, whether the cost of finding successful attacks has a higher priority than a lower rate of nonfunctional files. Ultimately, if the goal is to evaluate how vulnerable the target model is to a single yet powerful transformation sequence, then GAME-UP can provide an efficient response finding universal attacks.

Towards Robustness

11

She stood in the storm and when the wind did not blow her way, she adjusted her sails.
—Elizabeth Edwards

Having shown that ML models are susceptible to adversarial examples, we now explore potential strategies to harden malware classifiers, as displayed under *Defenses* in Fig. 1.1. In general, the success metric evaluated is based on minimizing the FNR, that is, the number of adversarial examples that effectively bypass the classifier. By measuring the baseline evasion rate and comparing it to the hardened model, we can establish the success of the defense mechanism implemented.

Before we begin with potential defense mechanisms, let us first briefly analyze the impact of normalization on malware classifiers trained with Windows PE files, since this technique is widely used when working with ML across multiple domains and may influence the performance of models before hardening.

Next, we investigate the effect of feature reduction as a potential defense mechanism by choosing a smaller set of ranked features, which report the highest values of importance. The goal of this strategy is to limit the number of features used by the model and, thus, decrease the sensitivity of the classifier when changes are applied to the input object. This also helps restrict the surface of attack for the adversary limiting their capabilities.

Then, we present UAP-based adversarial training, which aims to improve the resilience of models against such attacks, and introduce our approach, which is a byproduct of the GAME-UP module presented in Chapter 7 and extends traditional adversarial training [Sze+13]. In this section, we analyze how to leverage problem-space knowledge into improving models. Hence, we measure the effectiveness of hardening strategies to study how adversarial attacks contribute to the improvement of the resilience of malware classifiers against adaptive adversaries.

© The Author(s), under exclusive license to Springer Fachmedien Wiesbaden GmbH, part of Springer Nature 2023
R. Labaca-Castro, *Machine Learning under Malware Attack*,
https://doi.org/10.1007/978-3-658-40442-0_11

Because of the high diversity in the feature space of Windows PE files (i.e., values differing with several orders of magnitude) and observing the work of similar approaches in the image processing domain [Xia+18], applying scaling transformations to features can become an attractive approach to leverage low-cost improvements in terms of resilience. In general, neural networks can benefit from these techniques, since normalizing values generally transfer to modifications in the step size of the algorithm and potentially lead to updates in gradient descent at different rates.

With our settings, however, we target highly efficient decision tree models (i.e., LightGBM) in the problem space, which are not necessarily affected by normalization, since it only adjusts the optimal split of the tree branches without really impacting predictions and errors. Therefore, we proceed to explore defense techniques initially without scaling the dataset.

11.1   Feature Reduction

Selecting features can have positive impacts on a model, including alleviating the computational cost and eventually improving the performance. Therefore, we implement feature reduction to strengthen the detection of FNRs when the model is confronted with adversarial examples.

By transferring information from real attacks in the problem space, generated with actual PE files, to the feature space where only representations are analyzed, interesting information is obtained about the distribution of the features affected by transformations. As observed in § 7.4, certain features experience the highest delta variation and are regarded as indicators of evasiveness with respect to the model. Therefore, we explore the impact of reducing the domain space and retraining the classifier with a fraction of the highest modified features, since earlier studies have already reported the use of feature selection to evaluate the defenses of Android malware classifiers [ZD16]. Moreover, we exclude from this analysis similar alternatives, such as feature reduction through mutual information, following results previously reported in the context of malware evasion [Gro+16].

To extract the importance of features based on coefficient values and avoid using the target classifier to reduce bias, we train a separate Logistic Regression (LR) model using the EMBER dataset (§ 3.3.2), which reports an AUC-ROC score of 0.94.

Once the LR model is trained and the features are ranked, we follow the steps outlined in an earlier study [Fan+19] and reduce the input space to 20% of the most important features and extract them from the total to create a new training subset domain. By reducing the size of the domain, we expect the model to focus on learning less complex relationships as a result of the more limited number of features. Similarly, the impact of transformations may be less sparse and simpler to detect, yet at the risk of attributing a stronger impact on individual features (Figure 11.1).

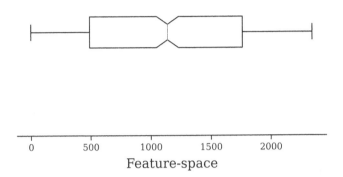

Feature-space

Figure 11.1 Distribution of the top 20% most important features according to the LR model

Next, we use the subset of features to retrain the model and observe whether it experiences a performance impact, since the original LightGBM has been trained without feature reduction.

In our experiments, we created a holdout test set with 200 malware examples and observed that when using only one-fifth of the features, the retrained model successfully flagged 70% more adversarial examples than the original classifier while still detecting 95% of the previous malware. This process seems to exhibit certain resilience toward adversarial malware examples, yet at the expense of the TPR. Therefore, the initial results indicate that feature reduction may be interesting to further explore in the context of Windows PE malware. However, we recommend deeper analyses in future studies and only focus here on more effective mechanisms.

11.2   UAP-Based Adversarial Training

From the shortcomings explored in § 11.1 and § 11.2 and inspired by the work on adversarial training [Sze+13], we leverage the knowledge acquired from the GAME-UP module (§ 7.4) to systematically identify weaknesses in classifiers in terms of universal adversarial attacks. Then, we evaluate the effectiveness of a hardened model against adaptive adversaries.

Training a model with adversarial examples has shown promising results, by improving the resilience of hardened classifiers [GSS15; KGB16; Mad+17]. The goal of this process is to generate a set of adversarial examples and use them as input objects by channeling them back to the model. This allows the decision boundary to include gaps close to authentic objects during training. This improvement, however, generates an imbalance. By covering more adversarial regions, the model risks missing legitimate input areas. Therefore, it is important to carefully establish a trade-off between adversarial and legitimate classes. Whenever the model is strongly robust, the TPR is affected. In contrast, if a legitimate malicious class is highly detected, then adding light perturbations is probably enough to generate adversarial examples.

Overall, improving the resilience of classifiers is associated with a number of challenges. In real-world attacks, fitting a classifier in the feature space does not necessarily match a real object in the problem space. In other words, the regions close to legitimate objects in terms of features may never represent a real adversarial malware example.

Moreover, in the context of Windows malware, in which files are parsed using libraries such as LIEF [Tho18], adversarial training should be implemented carefully. These libraries facilitate access to the PE file and its compilation after the injection of perturbations. However, this process may lead to idiosyncratic information being added to the file during adversarial training. In this case, the model may learn how to protect against *modified* and *not modified* objects instead of learning to classify between *malicious* and *benign*.

Another important aspect is that binary malware classification, unlike regular binary classification problems, aims to protect one class against adversarial attacks. This means that malware should be prevented to be detected as benign software, since adversaries are weakly motivated to generate adversarial benign examples[1]. A similar phenomenon occurs in multiclass problems, frequently observed in image processing, in which every class needs to be protected against adversarial examples.

This, however, does not suggest that adversarial training needs to be performed exclusively using adversarial examples, since, particularly for linear classifiers, this would considerably lead to an increase in bias toward the malware class. At this point, the features associated with benign software for baseline models can be indicative of malicious files for classifiers trained with adversarial examples. It may also

---

[1] Despite being interesting, injecting maliciousness into a, for example, *notepad.exe* file does not yield the same reward for the adversary compared to modifying a *ransomware* to make it undetectable, since the former is likely to be flagged by classifiers, whereas the latter is not.

impact the TPR as the model no longer focuses on legitimate malware but rather on adversarial input.

We therefore interpolate legitimate malicious files with adversarial malware at a 50% ratio. This process is called *mixed* adversarial training. In contrast, *pure* adversarial training is the case in which adversarial examples are generated only for retraining the baseline model. We can broadly define the process of our UAP-based adversarial training in the following steps:

i) Generate realizable adversarial examples in the problem space by searching greedily through the *exploration set* to identify UAP candidates using a toolkit of transformations (§ 3.3.3). This stage involves calculating the initial robustness level of the baseline model.

ii) Identify potential universal sequences by evaluating all UAP candidates using the *validation set* set and selecting as UAP the sequence of transformations that maximizes the evasion rate across the input objects.

iii) Extract UAP noise by deriving statistics from the distribution of problem-space UAPs transferred into the feature space, which maps the impact of UAP byte-level perturbations to the features.

iv) Produce a set of synthetic adversarial examples using the derived statistics, which approximate the distribution of real adversarial malware examples to make the process more efficient.

v) Train the model by employing adversarial data either with a *pure* approach using the adversarial examples derived from the statistics or with a *mixed* approach combining the malware set with the synthetic generated examples.

vi) Evaluate the resilient models, once they have been adversarially trained, by performing an attack and searching for new UAP candidates (§ 7.3). In this step, the models are measured using the UER metric in terms of the TPR cost on legitimate malware examples.

11.2.1 Evaluate Hardened Models

To identify systemic weaknesses in a model, effective adversarial examples need to be generated. However, producing problem-space adversarial examples in the context of malware is considerably costlier than in the feature space. Additionally, the LightGBM is not trained in batches through a number of epochs as neural networks are, increasing the cost to harden the model.

We address this limitation by deriving statistics from the UAP distribution in the feature space when induced by transformations in the problem-space toolkit. In essence, we sample random transformations through a statistical model to produce adversarial examples in the feature space, which significantly reduces the number of problem-space examples required. Although related approaches show that training with *unrealistic examples* (synthetic data) using PE files does not improve the resilience for input-specific attacks [Dyr+22], our derived statistics show positive progress in the UAP domain. In addition, while this process does not consider the relationship between malware features, which may not always be the case, our empirical experiments show that it can harden models successfully against adaptive adversaries using UAP attacks.

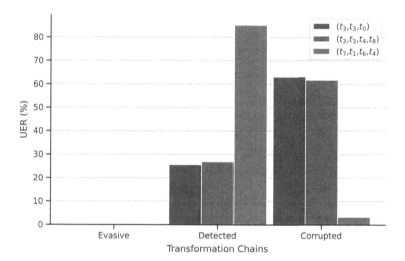

Figure 11.2 New UAP candidates (t_3, t_3, t_0) and (t_3, t_3, t_4, t_8), along with previous UAP (t_7, t_1, t_6, t_4), are not evasive in the hardened model bringing the UER close to zero, while significantly increasing detected and corrupted candidates

For our evaluation, we consider two confidence thresholds: $T = .90$, which is commonly used in previous work [And+18; AR18], and $T = .87$, which matches 0.1% FPR in our experimental settings.

Since pure adversarial training incurs a significant cost for legitimate detection performance decreasing the AUC-ROC of the model $(T = .87)$ to .624, and considering that the mixed approach maintains the performance at .853, we focus on the latter for our experiments, as shown in Table 11.1.

We follow the best practices and execute a new attack to evaluate the retrained model's resilience. Overall, the newly calculated UAP candidates (t_3, t_3, t_0) and

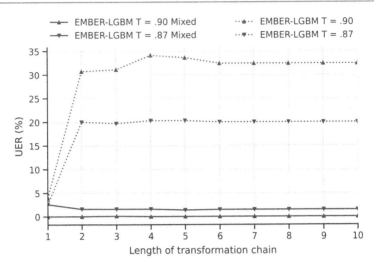

Figure 11.3 Adaptive adversary attacks against a LightGBM classifier, showing increasing UER values at diverse sequence lengths. Conversely, adversarial training using mixed strategy shows promising results for both confidence scores ($T = .87$ and $T = .90$) causing the UER to be close to zero

(t_3, t_3, t_4, t_8) differ from the previous UAP (t_7, t_1, t_6, t_4) (§ 7.4). However, they are still correctly detected by the hardened model while generating a large number of nonfunctional PE files. The results are shown in Fig. 11.2. In fact, the model becomes robust not only against the previous UAP, but also at correctly classifying current false negatives. Fig. 11.3 shows that 99.8% of the adversarial examples are successfully detected. Therefore, it can be observed that UAP-based adversarial training using mixed strategies is a promising approach to improve the detection of universal attacks, as it raises the complexity for the adversary to identify new vulnerabilities successfully.

11.2.2 UAP Search Alternative

After analyzing our UAP attack using a greedy strategy and having collected promising results with AIMED using GP, we experimented combining UAP attacks with evolutionary algorithms. The underlying idea is to generate faster universal attacks that could be used to improve defenses more efficiently since greedy algorithms require considerable search time.

Therefore, we evaluate GP as an alternative UAP search strategy. However, given that GP assigns more importance to the sequences rather than individual perturbations, the UER is often negatively impacted. In this scenario, while multiple UAPs are successfully identified, GP returns higher fitness functions to repetitive sequences, which leads to an increase in integrity issues with the PE files and, hence, reduces the plausibility of the binary.

Unlike input-specific attacks, in which GP can converge by considering *evasion* and *integrity*, searching for UAPs is a more complex problem, in which the algorithm needs to find a single sequence of transformations that optimizes evasion across a large fraction of objects while preserving the plausibility of the adversarial examples. It is also worth highlighting that the attack exhibits some issues pertaining to successful generalization across multiple input objects. As shown in Table 11.1, the approach reports low UERs for both confidence thresholds of the LightGBM classifier: $T = 0.90$ and $T = 0.87$.

Compared to the results obtained using the *greedy* approach (§ 7.3), in which the UERs reported reach up to 30%, we observe that GP does not yield improvements when searching for universal attacks achieving an evasion rate of less than 4%. In fact, despite being very efficient in identifying input-specific attacks (§ 5.3), GP algorithms still require heavy improvements to be competitive when searching for successful UAP sequences to generate adversarial malware examples in the problem space. Thus, we continue to rely on greedy approaches to harden malware classifiers until better strategies are found.

Table 11.1 Comparison between the baseline ML-based model and our problem-space defenses against an adaptive attacker (UER at |T| of 1, 4, and 10) using greedy approach (GR) and GP as the UAP search strategy. The models targeted implement thresholds from literature (.90) and at 0.1% FPR (.87)

	MODEL	SEARCH	AUC- ROC	TPR	UER_1	UER_4	UER_{10}
Baseline	LightGBM $T = .90$	GR	0.999	0.921	4.0%	34.1%	32.4%
	LightGBM $T = .87$	GR	0.999	0.930	2.6%	20.3%	20.0%
	LightGBM $C = .90$	GP	0.999	0.921	—%	0.6%	3.5%
	LightGBM $C = .87$	GP	0.999	0.930	—%	0.0%	0.1%
Defended	LightGBM $T = .90$	—	0.988	0.836	0.0%	0.1%	0.01%
	LightGBM $T = .87$	—	0.998	0.853	2.6%	1.6%	1.5%

11.3 Summary

In summary, universal adversarial attacks are a risk for models and need to be properly addressed since they represent a critical issue compared with input-specific attacks. UAP-derived defenses can significantly improve the resilience of ML-based malware classifiers against adaptive adversaries, increasing the cost to generate fresh effective attacks. Therefore, our proposed defense shows that a variant of adversarial training using a combination of real and synthetic objects substantially improves the resilience against universal attacks.

We evaluated the results of UAP attacks using two search strategies, greedy and GP. As shown in Table 11.1 and discussed in § 11.2.2, GP significantly underperforms compared to the greedy approach. Moreover, while GP has been shown to be successful in generating efficient input-specific attacks (§ 5.2), the results obtained when searching for UAPs are less effective. Even though the GP approach can be further optimized for better results, we believe that additionally strategies can be explored to improve the efficiency of hardening malware models, because of the computational requirements of greedy algorithms. In terms of protecting classifiers against universal attacks, our UAP-based defense can reportedly avert UAP attacks with minor impact on performance and very high detection rates.

Part VI
Closing Remarks

Conclusions & Outlook

12

No book can ever be finished. While working on it we learn just enough to find it immature the moment we turn away from it.
—Karl Popper

In this chapter, we conclude our work by providing a brief summary of our research and the main contributions presented. Next, we reflect on the lessons learned by connecting the dots and provide potential directions for future work.

In this dissertation, we explored the weaknesses of ML-based static malware classifiers against adversarial attacks. We considered the context of Windows malware, in which PE files are subject to byte-level modifications during test time. We introduced multiple attack strategies using stochastic methods and learning algorithms to enhance the search for optimal attacks.

Moreover, we studied the relationship between the length and evasiveness of sequences to understand the impact of larger attacks compared to minimal perturbations. As a result, we found that using a toolbox of semantic-preserving transformations does not guarantee valid PE files. Therefore, we added an integrity verification stage of adversarial examples to ensure that attacks are realizable in the problem space at a reasonable cost. Hence, our approach shows that it is generally possible to produce valid binaries when targeting ML-based malware classifiers.

We finally analyzed different strategies to improve the resilience of classifiers against adaptive adversaries on the basis of the problem-space knowledge acquired by generating universal adversarial malware examples.

© The Author(s), under exclusive license to Springer Fachmedien Wiesbaden GmbH, part of Springer Nature 2023
R. Labaca-Castro, *Machine Learning under Malware Attack*,
https://doi.org/10.1007/978-3-658-40442-0_12

12.1   Revising Research Questions

In this section, we address the research questions presented in Chapter 1 and provide responses regarding the effects of adversarial examples on ML malware classifiers.

RQ1: How can we define a systematic approach that allows weaknesses of malware models to be methodologically assessed using real-world attacks?

In Chapter 3, we introduced a scheme to systematically evaluate ML-based static malware classifiers for Windows PE files. The underlying goal was to automate the exploration of vulnerabilities in malware models using more realistic attack scenarios in the problem space, as opposed to unrealizable attacks in the feature domain. With that in mind, we started by defining the theoretical framework (§ 3.1) for adversarial attacks in both the feature space and the problem space along with their constraints. Although the focus remained on the latter, we specified both domain spaces, since attacks are supported and introduced in the two settings.

For the threat model in § 3.2, we presented the adversary's goals, knowledge, and capabilities. We explore PK scenarios, in which adversaries can retrieve gradient information directly and may also know the training set, and the constraints of LK attacks that reduce the ability to produce effective adversarial examples.

Next, we outlined the settings with target models (§ 3.3.1) and datasets (§ 3.3.2) that are accessible to encourage reproducibility and improvements in the performance of evasion attacks with realizable PE binaries. Likewise, the toolbox of byte-level transformations (§ 3.3.3) can be unlimitedly extended to maximize the effectiveness of adversarial attacks.

We additionally implemented an integrity verification stage in § 3.3.4 to ensure that the adversarial malware examples preserve plausibility throughout the pipeline and can, thus, be later reused to adversarially train the target models. While relevant studies in the literature have rejected the use of verification steps for efficiency reasons, we argue that (highly) evasive adversarial examples are only useful when practical. Therefore, we optimized the system to provide realizable attacks at a reasonable cost without relying exclusively on functionality-preserving PE transformations.

RQ2: To what extent can the generation of adversarial examples against malware classifiers be efficiently automated while remaining plausible?

We address this question by breaking it down into *plausibility* and *efficiency*, since both aspects require individual consideration.

First, in terms of *plausibility*, we explored in Chapter 4 the impact of byte-level perturbations when injected into Windows binaries. Given that PE files typically provide no access to the source code, transformations require manipulating the structure of the file at the byte level, which often leads to compromising the plausibility of the file. Therefore, we showed that *semantic preserving* perturbations can still negatively influence adversarial examples, making a case for the use of integrity verification stages throughout the process. Moreover, since the nature of PE files increases the complexity of modifying objects in the context of malware, approaches that dismiss verification are obliged to significantly reduce the toolbox to a a small set of one or two transformations. This allows the attack to become simpler to detect because of the lack of diversity.

Counterintuitively, larger sequences of transformations are not required to achieve greater levels of evasion. In fact, experiments have shown that, in certain cases, sequences of 500 transformations reach the same evasion rates as vectors of five, emphasizing that highly modified files are not only complex to generate, but also more detectable by malware classifiers (Table 4.1).

In contrast, as expected, larger sequences tend to make the work of finding valid examples more difficult (Fig. 4.2). We thereby showed that having *four* and *six* perturbations is already enough to secure an effective rate of evasiveness (Fig. 4.3) while retaining reasonable levels of plausibility.

Second, for *efficiency* reasons, after introducing the ARMED module, which automates the generation of valid adversarial malware examples, we shifted our focus toward improving the performance of attacks by exploring multiple strategies.

We started by introducing AIMED, which implements a GP approach to search for optimal sequences of transformations against malware classifiers. GP significantly improves the performance of traditional stochastic models by reducing the time needed to converge and find adversarial examples to half, as shown in Fig. 5.3(a), and also by producing less than a fourth of nonfunctional malware objects as depicted in Fig. 5.3(b).

While GP offers many benefits compared to traditional approaches, it is also known for falling into local minima, which compromises the search for optimal results. Thus, we explored further techniques, such as RL. In Chapter 6, we introduced AIMED-RL, whose goal is to search for effective attacks using RL agents. As shown in Table 6.4, our approach provides the best trade-off between evasion rates and the number of perturbations required. Our RL-based attack also shows almost the same number of adversarial examples than similar work by using only ∼6% of the perturbations required by the most evasive approaches. It also addresses the problem of overutilizing the same transformation by introducing a *penalty technique*, which leads to more diverse attacks as reported in Fig. 6.2.

To summarize, we compared our attack modules and report the results in Table 10.1, in which we observe that there is no single approach that fits all cases. Each strategy favors a certain scenario. Moreover, while random transformations yield effective results when searching for single attacks, optimized solutions seem to be more efficient when scaling the attack to systematically query the models using a larger set of adversarial examples. Finally, both GP and RL show similar evasion rates ($\sim$42%). Although the former searches for adversarial examples heuristically and the latter requires training the agents, RL exhibits a great functionality rate with 97.6% valid adversarial examples.

RQ3: Are there effective universal attacks in the context of malware and how can these enhance a model's resilience against adaptive adversaries?

In Chapter 7, we introduced universal attacks in the problem space in the context of malware. We showed that realizable UAPs are feasible for malware files and that the evasion rates are competitive when compared against input-space attacks. The results are shown in Table 10.1. Overall, UAPs allow computing a single sequence of transformations that can then be applied to a large fraction of objects regardless of the malware nature (e.g., size, type, and family). Additionally, we showed that a successful UAP does not significantly impact the plausibility of the file. Fig. 7.1 shows the UERs of UAP candidates with the most successful sequence achieving 30% evasion. Interestingly, the order of the sequence indicates a solid success factor, since appending t_3 on (t_7, t_1) substantially affects the outcome of the adversarial example and reduces the level of plausibility, hence increasing the nonfunctional rates from $\sim$5% to 29%. In contrast, appending t_6 continues to improve the candidate's success, hence, increasing the overall UER.

We evaluated our experiments in the problem space by analyzing the delta variation for each UAP in the feature space. As shown in Fig. 7.2, we verified that all candidates present a similar distribution throughout their features, which further supports the evasiveness against the target model. We also showed that sequences with reduced functionality rates present a clearly distinct pattern compared to valid adversarial malware examples.

In terms of resilience, we leveraged the knowledge from universal perturbations in the problem space to improve the classifier's ability to detect adversarial examples. We introduced UAP-based adversarial training that relies on using a mixture of legitimate and adversarial examples as malicious labeled malware to retrain the model. Our experiments showed that the ability to find new UAP candidates against the hardened model is significantly reduced, causing the evasion rate to be close to zero, as shown in Fig 11.2. Moreover, we verified that the previous UAP sequence against the baseline model is successfully detected by the hardened model. Hence, we confirmed the performance of the defense approach. The UER of the baseline and hardened LightGBM models are shown in Fig. 11.3.

12.2 Contributions

In this section, we summarize the main contributions of this work. We start by addressing the empirical guidance in the context of malware, in which multiple approaches are introduced to identify weaknesses in malware classifiers. We then review our proposed UAP defense to mitigate powerful universal attacks and release the framework to encourage further research.

12.2.1 Empirical Guidance for Malware Attacks

In this work, we introduced multiple attacks in the problem space for Windows PE files that produce realizable adversarial malware examples against ML-based classifiers. We analyzed the nuances of diverse approaches and explored the best applications for each scenario, from the generation of individual attacks to the identification of systemic weaknesses in malware models. We also highlighted the need for integrity verification stages, given that even semantic-preserving perturbations may have a negative impact on the plausibility of objects. We conducted our attacks by targeting both individual models directly and aggregator platforms with over 60 classifiers simultaneously. We also confirmed cross-evasion between models including research and commercial classifiers.

By mainly generating realizable attacks, we drew a comparison of Windows malware evasion targeting static classifiers, which can be used as a benchmark for future research on PE malware. Unlike in the computer vision domain, in which the existing measures are well known, clear evasion rate comparisons are missing in the context of malware, and each approach seems to work virtually isolated.

Finally, we explored feature-space attacks using real-world perturbations and gradient information. Despite implementing less realistic settings compared to full problem-space approaches, attacks in the feature space can help better understand the weaknesses of models against adversarial malware examples, since full visibility is provided.

12.2.2 Adversarial Defenses derived from UAP attacks

To the best of our knowledge, we have introduced the first realizable UAP attack and defense against universal attacks in the context of malware. We showed how the knowledge extracted from UAP attacks can be used to improve the resilience of malware classifiers, rendering fresh universal attacks ineffective at a reasonable cost in terms of TPRs. Besides, we trained the hardened model to cover the pockets of vulnerabilities, leading to the correct detection of previously successful UAPs, which demonstrates effective defenses against such attacks.

While the robustness of ML models against adaptive adversaries remains an open problem across multiple domains, we believe that this is a step in the right direction for malware classifiers, since universal attacks can be further explored by adversaries to improve their capabilities. Furthermore, UAP attacks reduce the cost for adaptive adversaries to find new attack vectors with close to zero impact on the plausibility of the adversarial examples, as opposed to input-specific attacks, in which the effectiveness depends on the methodology chosen.

Therefore, the main goal of these adversarial defenses is to increase the cost for the adversaries to compute new universal adversarial examples, demanding them to either generate a new toolset of problem-space transformations or further rely on input-specific attacks. However, both options are expensive and exhibit considerable challenges. For example, for the former, creating new realizable perturbations that are strong enough to evade classifiers yet subtle enough to keep the files valid is not a trivial process and demands specific expertise. On the other hand, the latter option requires computing new attacks every time an input object is provided, which increases the prospect of affecting the integrity of the PE files.

12.2.3 Framework to Evaluate Malware Classifiers

In this work, we implemented the aforementioned attacks and defenses in FAME and released it as an open-source[1] [Lab22] modular framework, under FAIR principles[2], to evaluate adversarial malware examples in the Windows domain. While this framework was initially designed to support PE files, further extensions can allow it to support further platforms. We now summarize the three most important aspects of our framework. First, its modular nature allows easy-to-implement deve-

---

[1] https://github.com/zRapha/FAME

[2] The FAIR guiding principles for scientific data management state that research objects should be Findable, Accessible, Interoperable, and Reusable [Wil+16].

lopment. The toolbox with problem-space perturbations can be extended without further knowledge of the attack strategy that is implementing it. Conversely, new attack modules can be designed without the need for a deep analysis of the toolbox. Second, it allows evaluating the impact of combining perturbations in optimized ways using multiple strategies. It also shows the limitations of byte-level perturbations in the context of malware. For example, semantic-preserving transformations can lead to corrupt adversarial examples if large sequences or specific variations are adopted. Third, it implements a suite of attacks to assess ML-based models, including stochastic, GP, RL, and universal approaches, to generate realizable adversarial malware examples.

Finally, attack modules can be used independently to address the issues of models. A defense module is also provided to increase the resilience of malware classifiers.

12.3   Connecting the dots

In this section, our goal is to connect the dots by investigating the process of working with AML in the context of malware and to share a few lessons learned.

Collecting Benign Data

For some domains, collecting data is a challenge in ML research, especially when working with datasets that can have security implications. As expected, this is the case in the context of malware, since companies owning such datasets avoid granting public access to them to prevent data leaks. Although in the past acquiring malware files required collaboration with commercial players and complicated nondisclosure agreements, the situation is now different. Malware repositories are nowadays publicly available and the level of accessibility has tremendously increased in the last few years.

However, the problem now is to find benign data. Although some might think that finding clean files should be a simple endeavor, this remains an underestimated challenge. Given the licensing constraints, most organizations working with benign data avoid making their datasets available to prevent legal disputes. This increases the cost of training malware models. While standardized datasets do exist, they are still scarce, which paves the way for the next point.

Scarce Datasets for Malware

Overall, some datasets containing the features with which ML models can be trained are available. However, the options are still very limited for Windows binaries. While computer vision applications can leverage more static training data from a variety of sets, including MNIST, ImageNet, and Fashion MNIST, malware datasets need to be constantly updated to include new threats and eventually prevent drift over time.

Such a limitation poses a challenge for research, since classifiers require updated data that are often not available. In some cases, new datasets are modified with deactivated features before being published to prevent malicious spread. However, despite being reasonable in terms of security, this may bias models toward less realistic representations of malware binaries.

Benchmarking Malware Classifiers
Another limitation is the availability of benchmarks comparing different approaches for adversarial evasion in the context of malware, especially for the Windows platform. Since generating real problem-space adversarial examples targeting black- and gray-box classifiers is a more complex task for Windows than for other platforms (e.g., Android APKs or PDF files), the results obtained are modest. In some cases, adversarial examples in Android report 100% evasion, whereas Windows PEs in the problem space are not even half as powerful. As discussed throughout this work, this behavior has multiple reasons. However, such a disparity in complexity and in the results presents a hurdle when publishing in the field. In other words, from the ML perspective, research in the context of PE malware is inevitably compared to other domains, including computer vision, in which adversarial examples in the problem space are invertible and report results close to perfect evasion. Such a behavior confuses the complexity of PE malware with lack of merit and therefore prevents more research in the field.

12.4   Future Work

In this section, we analyze the potential lines of future research related to AML in the context of malware. Although there are many open problems in the field, we believe that addressing them is paramount to ensure that ML models are safe in their deployment, especially for malware classifiers, in which adversaries have strong motivations to identify effective attacks.

Automating Dynamic Evasions
Generally, research involving dynamic evasion based on malware behavior is considered an emerging field within Windows binaries. As explored throughout this work, most of the approaches focus on static malware classification, yet malware classifiers implement different layers of protection, including both static and dyna-

mic analyses of features. However, creating an automated pipeline that can influence malware behavior is not a trivial challenge. Moreover, when behavior is affected and evasion is eventually achieved, it is important to determine whether maliciousness is retained. One of the main directions to achieve adversarial malware is to imitate benign behavior. This, however, raises the question of what the difference would be between benign behavior imitation and *pure* benign behavior. In other words, there is no point in achieving misclassification if the adversarial example is no longer considered malicious.

Accessibility of Commercial Classifiers
A large portion of the literature focuses largely on research models, since industry classifiers are less accessible. Therefore, more efficient adaptations of commercial classifiers suited for research purposes are needed. Currently, this challenge can be solved via aggregators, which is an expensive option and requires academic APIs that are not typically available. Furthermore, licensing and legal limitations are considered a hurdle in performing assessments of commercial classifiers.

Improving Byte-level Perturbations
In general, more research into creating a toolbox of transformations, especially for Windows PE files, is required. Creating effective binary transformations is a challenging issue. New attempts largely rely on appending bytes at the ends of sections or the ends of files. Despite being effective, these are simple to identify during preanalyses and do not pose a large risk for real attacks and are, thus, less effective in evaluating the weaknesses of malware classifiers.

Focusing on Robustness of Malware Classifiers
Finally, while robustness is currently an active field of research, more approaches focusing on the intricacies of malware classifiers in the real world are required. For instance, in the context of malware, valid adversarial examples in the problem space are costly to generate, which limits the impact of adversarial training techniques. Therefore, it would be interesting to explore more effective strategies to produce real adversarial binaries that can be used to harden Windows malware classifiers. Further techniques beyond our domain may also be interesting to further explore, such as leveraging knowledge from differential privacy.

List of Publications

FAME: Framework for Adversarial Malware Evaluation. In: *Journal currently under review*, 2022. **Raphael Labaca-Castro** and Gabi Dreo Rodosek.
Realizable Universal Adversarial Perturbations for Malware. In: *ArXiv arXiv: 2102.06747*. 2022. **Raphael Labaca-Castro**. Luis Muñoz-González, Feargus Pendlebury, Gabi Dreo Rodosek, Fabio Pierazzi, and Lorenzo Cavallaro.
Attacking malware classifiers by crafting gradient-attacks that preserve functionality. In: *Proceedings of the 2019 ACM SIGSAC Conference on Computer and Communications Security (CCS). Poster article*. 2019, 2565–2567. **Raphael Labaca-Castro**, Battista Biggio, and Gabi Dreo Rodosek.
AIMED-RL: Exploring Adversarial Malware Examples with Reinforcement Learning. In: *Joint European Conference on Machine Learning and Knowledge Discovery in Databases (ECML-PKDD). Springer Editorial*. 2021, 37–52. **Raphael Labaca-Castro**, Sebastian Franz, and Gabi Dreo Rodosek.
AIMED: Evolving Malware with Genetic Programming to Evade Detection. In: *2019 18th IEEE International Conference on Trust, Security and Privacy in Computing and Communications (TRUSTCOM). IEEE*. 2019, 240–247. **Raphael Labaca-Castro**, Corinna Schmitt, and Gabi Dreo Rodosek.
ARMED: How Automatic Modifications Can Evade Detection? In: *2019 5th IEEE International Conference on Information Management (ICIM)*. 2019, 20–27. **Raphael Labaca-Castro**, Corinna Schmitt, and Gabi Dreo Rodosek.
IoT-botnet detection and isolation by access routers. *2018 9th IEEE International Conference on the Network of the Future (NOF)* (2018). Christian Dietz, **Raphael Labaca-Castro**, Jessica Steinberger, Cezary Wilczak, Marcel Antzek, Anna Sperotto, and Aiko Pras.
OpenMTD: A Framework for Efficient Network-Level MTD Evaluation. *2020 27th ACM Moving Target Defense Workshop co-located with Conference on Computer and Communications Security (MTD@CCS)*, (2020). Richard Poschinger, Nils Rodday, **Raphael Labaca-Castro**, and Gabi Dreo Rodosek.

R. Labaca-Castro, *Machine Learning under Malware Attack*, https://doi.org/10.1007/978-3-658-40442-0

References

[Abd+19]   Sajjad Abdoli, Luiz G Hafemann, Jerome Rony, Ismail Ben Ayed, Patrick Car-
            dinal, and Alessandro L Koerich. **Universal adversarial audio perturbations.**
            arXiv preprint arXiv:1908.03173 (2019).
[AC18]   Giovanni Apruzzese and Michele Colajanni. **Evading botnet detectors based
            on flows and random forest with adversarial samples.** In: 2018 *IEEE 17th
            International Symposium on Network Computing and Applications (NCA).*
            IEEE. 2018, 1–8.
[AE20]   Abdullah Ali and Birhanu Eshete. **Best-Effort Adversarial Approximation of
            Black-Box Malware Classifiers.** In: *International Conference on Security and
            Privacy in Communication Systems.* Springer. 2020, 318–338.
[Al-+18]   Abdullah Al-Dujaili, Alex Huang, Erik Hemberg, and Una-May O?Reilly.
            Adversarial deep learning for robust detection of binary encoded malware.
            In: 2018 *IEEE Security and Privacy Workshops (SPW).* IEEE. 2018, 76–82.
[And+18]   Hyrum S. Anderson, Anant Kharkar, Bobby Filar, David Evans, and Phil Roth.
            Learning to Evade Static PE Machine Learning Malware Models via RL.
            ArXiv (2018). URL: http://arxiv.org/pdf/1801.08917v2.
[And+19]   Ross Anderson, Chris Barton, Rainer Böhme, Richard Clayton, Carlos Ganan,
            Tom Grasso, Michael Levi, Tyler Moore, and Marie Vasek. **Measuring the
            changing cost of cybercrime** (2019).
[AP95]   M Abrams and Harold J Podell. **Malicious software.** *Information Security*
            (1995).
[Apr+20]   Giovanni Apruzzese, Mauro Andreolini, Mirco Marchetti, Andrea Venturi, and
            Michele Colajanni. **Deep reinforcement adversarial learning against botnet
            evasion attacks.** *IEEE Transactions on Network and Service Management* 17:4
            (2020), 1975–1987.
[AR18]   Hyrum S. Anderson and Phil Roth. **EMBER: An Open Dataset for Training
            Static PE Malware Machine Learning Models.** *ArXiv* (2018). URL: https://
            arxiv.org/pdf/1804.04637.
[Arp+14]   Daniel Arp, Michael Spreitzenbarth, Malte Hubner, Hugo Gascon, Konrad
            Rieck, and CERT Siemens. **Drebin: Effective and explainable detection of
            android malware in your pocket.** In: *Ndss.* Vol. 14. 2014, 23–26.
[AS15]   E. Aydogan and S. Sen. **Automatic Generation of Mobile Malwares Using
            Genetic Programming.** In: *European Conference on the Applications of Evo-*

lutionary Computation. EvoApplications. Heidelberg, Germany: Springer, Mar. 2015, 745–756.

[Ath+18]   Anish Athalye, Logan Engstrom, Andrew Ilyas, and Kevin Kwok. **Synthesizing Robust Adversarial Examples**. In: *International Conference on Machine Learning.* 2018, 284–293.

[AV-20]   AV-TEST. **Security Report 2020** (2020).

[AWF16]   Hyrum S Anderson, Jonathan Woodbridge, and Bobby Filar. **DeepDGA: Adversarially-tuned domain generation and detection**. In: *Proceedings of the 2016 ACM Workshop on Artificial Intelligence and Security.* 2016, 13–21.

[Bar+06]   Marco Barreno, Blaine Nelson, Russell Sears, Anthony D Joseph, and J Doug Tygar. **Can Machine Learning be Secure?** In: *Procs. of the Symposium on Information, Computer and Communications Security.* 2006, 16–25.

[Bar+10]   Marco Barreno, Blaine Nelson, Anthony D Joseph, and J Doug Tygar. **The security of machine learning**. *Machine Learning* 81:2 (2010), 121–148.

[BDM17]   Marc G Bellemare, Will Dabney, and Rémi Munos. **A distributional perspective on reinforcement learning**. In: *International Conference on Machine Learning.* PMLR. 2017, 449–458.

[Ben12]   Yoshua Bengio. **Practical recommendations for gradient-based training of deep architectures**. In: *Neural networks: Tricks of the trade.* Springer, 2012, 437–478.

[BFR10]   Battista Biggio, Giorgio Fumera, and Fabio Roli. **Multiple classifier systems for robust classifier design in adversarial environments**. *International Journal of Machine Learning and Cybernetics* 1:1 (2010), 27–41.

[Big+13]   Battista Biggio, Igino Corona, Davide Maiorca, Blaine Nelson, Nedim Šrndić, Pavel Laskov, Giorgio Giacinto, and Fabio Roli. **Evasion attacks against machine learning at test time**. In: *Joint European conference on machine learning and knowledge discovery in databases.* Springer. 2013, 387–402.

[BNL12]   Battista Biggio, Blaine Nelson, and Pavel Laskov. **Poisoning attacks against support vector machines**. *arXiv preprint* arXiv:1206.6389 (2012).

[BR17]   Marie Baezner and Patrice Robin. **Stuxnet**. Tech. rep. ETH Zurich, 2017.

[BR18]   Battista Biggio and Fabio Roli. **Wild patterns: Ten years after the rise of adversarial machine learning**. *Pattern Recognition* 84 (2018), 317–331.

[Bro+16]   G. Brockman, V. Cheung, L. Pettersson, J. Schneider, J. Schulman, J. Tang, and W. Zaremba. **Openai Gym**. *arXiv preprint* arXiv:1606.01540 (June 2016), 1–4.

[Bro+17]   Tom B Brown, Dandelion Mané, Aurko Roy, Martin Abadi, and Justin Gilmer. **Adversarial Patch**. *arXiv preprint* arXiv:1712.09665 (2017).

[Bro+20]   Tom Brown, Benjamin Mann, Nick Ryder, Melanie Subbiah, Jared D Kaplan, Prafulla Dhariwal, Arvind Neelakantan, Pranav Shyam, Girish Sastry, Amanda Askell, et al. **Language models are few-shot learners**. *Advances in neural information processing systems* 33 (2020), 1877–1901.

[Cal+18]   Alejandro Calleja, Alejandro Martín, Héctor D Menéndez, Juan Tapiador, and David Clark. **Picking on the Family: Disrupting Android Malware Triage by Forcing Misclassification**. *Expert Systems with Applications* 95 (Apr. 2018), 113–126.

[Car+19]   Nicholas Carlini, Anish Athalye, Nicolas Papernot, Wieland Brendel, Jonas Rauber, Dimitris Tsipras, Ian J. Goodfellow, Aleksander Madry, and Alexey

Kurakin. **On Evaluating Adversarial Robustness**. *CoRR* abs/1902.06705 (2019).

[CGR13] Igino Corona, Giorgio Giacinto, and Fabio Roli. **Adversarial attacks against intrusion detection systems: Taxonomy, solutions and open issues**. *Information Sciences* 239 (2013), 201–225.

[Cha+18] Anirban Chakraborty, Manaar Alam, Vishal Dey, Anupam Chattopadhyay, and Debdeep Mukhopadhyay. **Adversarial attacks and defences: A survey**. *arXiv preprint* arXiv:1810.00069 (2018).

[Cho+19] Jusop Choi, Dongsoon Shin, Hyoungshick Kim, Jason Seotis, and Jin B Hong. **AMVG: Adaptive Malware Variant Generation Framework Using Machine Learning**. In: *2019 IEEE 24th Pacific Rim International Symposium on Dependable Computing (PRDC)*. IEEE. 2019, 246–24609.

[Cla+21] Claudio Guarnieri, Alessandro Tanasi, Jurriaan Bremer, and Mark Schloesser. **Cuckoo Sandbox—Automated Malware Analysis**. *Cuckoo* (2021). URL: https://cuckoosandbox.org (visited on 04/26/2021).

[Co+19] Kenneth T. Co, Luis Muñoz-González, Sixte de Maupeou, and Emil C. Lupu. **Procedural Noise Adversarial Examples for Black-Box Attacks on Deep Convolutional Networks**. In: *CCS*. ACM, 2019, 275–289.

[Coh86] Fred Cohen. **Computer viruses**. PhD thesis. University of Southern California California, 1986.

[Coh87] Fred Cohen. **Computer viruses: theory and experiments**. *Computers & security* 6:1 (1987), 22–35.

[Coh89] Fred Cohen. **Models of practical defenses against computer viruses**. *Computers & Security* 8:2 (1989), 149–160.

[CW19] Lena Y Connolly and David S Wall. **The rise of crypto-ransomware in a changing cybercrime landscape: Taxonomising countermeasures**. *Computers & Security* 87 (2019), 101568.

[Dal+04] Nilesh Dalvi, Pedro Domingos, Sumit Sanghai, and Deepak Verma. **Adversarial classification**. In: *Proceedings of the tenth ACM SIGKDD international conference on Knowledge discovery and data mining*. 2004, 99–108.

[Dem+17] Ambra Demontis, Marco Melis, Battista Biggio, Davide Maiorca, Daniel Arp, Konrad Rieck, Igino Corona, Giorgio Giacinto, and Fabio Roli. **Yes, machine learning can be more secure! a case study on android malware detection**. *IEEE Transactions on Dependable and Secure Computing* 16:4 (2017), 711–724.

[Dem+20] Luca Demetrio, Scott E Coull, Battista Biggio, Giovanni Lagorio, Alessandro Armando, and Fabio Roli. **Adversarial EXEmples: A Survey and Experimental Evaluation of Practical Attacks on Machine Learning for Windows Malware Detection**. *arXiv preprint* arXiv:2008.07125 (2020).

[Dem+21] Luca Demetrio, Battista Biggio, Giovanni Lagorio, Fabio Roli, and Alessandro Armando. **Functionality-preserving black-box optimization of adversarial windows malware**. *IEEE Transactions on Information Forensics and Security* 16 (2021), 3469–3478.

[Dep21] Statista Research Department. **Market share of leading computer operating systems worldwide**. *Statista* (2021). URL: https://www.statista.com/statistics/268237/global-market-share-held-by-operating-systems-since-2009/ (visited on 03/21/2022).

[Dyr+22]   Salijona Dyrmishi, Salah Ghamizi, Thibault Simonetto, Yves Le Traon, and
            Maxime Cordy. **On The Empirical Effectiveness of Unrealistic Adversarial**
            Hardening Against Realistic Adversarial Attacks. *arXiv preprint*

[Fan+19]   Z. Fang, J. Wang, B. Li, S. Wu, Y. Zhou, and H. Huang. **Evading Anti-Malware.**
            Engines With Deep Reinforcement Learning. *IEEE Access* 7 (2019), 48867–
            48879. https://doi.org/10.1109/ACCESS.2019.2908033.

[Fan+20]   Yong Fang, Yuetian Zeng, Beibei Li, Liang Liu, and Lei Zhang. **DeepDetectNet**
            vs RLAttackNet: An adversarial method to improve deep learningbased
            static malware detection model. *PLOS ONE* 15:4 (2020), e0231626. issn:
            1932–6203. https://doi.org/10.1371/journal.pone.0231626.

[FBS19]   Aurore Fass, Michael Backes, and Ben Stock. **Hidenoseek: Camouflaging**
            malicious javascript inbenign asts. In: *Proceedings of the 2019 ACM SIGSAC*
            Conference on Computer and Communications Security. 2019, 1899–1913.

[FN17]   Mohammad Reza Faghani and Uyen Trang Nugyen. **Modeling the**
            propagation of Trojan malware in online social networks. *arXiv*
            preprint arXiv:1708.00969 (2017).

[Fos+08]   Marc Fossi, Eric Johnson, Dean Turner, Trevor Mack, Joseph Blackbird, David
            McKinney, Mo King Low, Téo Adams, Marika Pauls Laucht, and Jesse Gough.
            Symantec report on the underground economy. *Symantec Corporation* 51
            (2008).

[GBC16]   Ian Goodfellow, Yoshua Bengio, and Aaron Courville. **Deep Learning.** MIT
            Press, 2016.

[Goo+14]   Ian J Goodfellow, Jean Pouget-Abadie, Mehdi Mirza, Bing Xu, David Warde-
            Farley, Sherjil Ozair, Aaron Courville, and Yoshua Bengio. **Generative adver-**
            sarial networks. *arXiv preprint* arXiv:1406.2661 (2014).

[GR06]   Amir Globerson and Sam Roweis. **Nightmare at test time: robust learning**
            by feature deletion. In: *Proceedings of the 23rd international conference on*
            Machine learning. 2006, 353–360.

[Gro+16]   Kathrin Grosse, Nicolas Papernot, Praveen Manoharan, Michael Backes, and
            Patrick McDaniel. **Adversarial perturbations against deep neural networks**
            for malware classification. *arXiv preprint* arXiv:1606.04435 (2016).

[GS18]   Barbara J. Grosz and Peter Stone. **A Century-Long Commitment to Assess-**
            ing Artificial Intelligence and Its Impact on Society. *Commun. ACM* 61:12
            (Nov. 2018), 68–73. ISSN: 0001–0782. https://doi.org/10.1145/3198470. URL:
            https://doi.org/10.1145/3198470.

[GSS15]   Ian J Goodfellow, Jonathon Shlens, and Christian Szegedy. **Explaining and**
            Harnessing Adversarial Examples. In: *ICLR.* 2015.

[Hex21]   Hex-Rays. **IDA Pro—Hex Rays.** *IDA* (2021). URL: https://www.hex-rays.com/
            products/ida/ (visited on 06/07/2021).

[Hou+20]   Ruitao Hou, Xiaoyu Xiang, Qixiang Zhang, Jiabao Liu, and Teng Huang. **Uni-**
            versal Adversarial Perturbations of Malware. In: *International Symposium*
            on Cyberspace Safety and Security. Springer. 2020, 9–19.

[HT17]   Weiwei Hu and Ying Tan. **Generating adversarial malware examples for**
            black-box attacks based on GAN. *arXiv preprint* arXiv:1702.05983 (2017).

[Hua+11]   Ling Huang, Anthony D Joseph, Blaine Nelson, Benjamin IP Rubinstein, and J Doug Tygar. **Adversarial Machine Learning**. In: *Procs. of the Workshop on Security and artificial intelligence*. 2011, 43–58.

[Ibi+19]   Olakunle Ibitoye, Rana Abou-Khamis, Ashraf Matrawy, and M Omair Shafiq. **The Threat of Adversarial Attacks on Machine Learning in Network Security-A Survey**. *arXiv preprint* arXiv:1911.02621 (2019).

[Jan+20]   Joel Janai, Fatma Güney, Aseem Behl, Andreas Geiger, et al. **Computer vision for autonomous vehicles: Problems, datasets and state of the art**. *Foundations and Trends®in Computer Graphics and Vision* 12:1–3 (2020), 1–308.

[Jin+21]   Beomjin Jin, Jusop Choi, Hyoungshick Kim, and Jin B Hong. **FUMVar: a practical framework for generating fully-working and unseen malware variants**. In: *Proceedings of the 36th Annual ACM Symposium on Applied Computing*. 2021, 1656–1663.

[KB14]   Diederik P Kingma and Jimmy Ba. **Adam: A method for stochastic optimization**. *arXiv preprint* arXiv:1412.6980 (2014).

[Ke+17]   Guolin Ke, Qi Meng, Thomas Finley, Taifeng Wang, Wei Chen, Weidong Ma, Qiwei Ye, and Tie-Yan Liu. **LightGBM: A Highly Efficient Gradient Boosting Decision Tree**. In: *Advances in Neural Information Processing Systems*. Ed. by I. Guyon, U. V. Luxburg, S. Bengio, H. Wallach, R. Fergus, S. Vishwanathan, and R. Garnett. Vol. 30. Curran Associates, Inc, 2017, 3146–3154. URL: https://proceedings.neurips.cc/paper/2017/file/6449f44a102fde848669bdd9eb6b76fa-Paper.pdf.

[Kep+95]   Jeffrey O Kephart, Gregory B Sorkin, William C Arnold, David M Chess, Gerald J Tesauro, Steve R White, and TJ Watson. **Biologically inspired defenses against computer viruses**. In: *IJCAI (1)*. 1995, 985–996.

[KGB16]   Alexey Kurakin, Ian J. Goodfellow, and Samy Bengio. **Adversarial Machine Learning at Scale**. *CoRR* abs/1611.01236 (2016).

[KK92]   John R Koza and John R Koza. **Genetic programming: on the programming of computers by means of natural selection**. Vol. 1. MIT press, 1992.

[KL93]   Michael Kearns and Ming Li. **Learning in the presence of malicious errors**. *SIAM Journal on Computing* 22:4 (1993), 807–837.

[KLA19]   Tero Karras, Samuli Laine, and Timo Aila. **A style-based generator architecture for generative adversarial networks**. In: *Proceedings of the IEEE/CVF conference on computer vision and pattern recognition*. 2019, 4401–4410.

[KO18]   Valentin Khrulkov and Ivan V. Oseledets. **Art of Singular Vectors and Universal Adversarial Perturbations**. In: *CVPR*. IEEE Computer Society, 2018, 8562–8570.

[Kol+18]   Bojan Kolosnjaji, Ambra Demontis, Battista Biggio, Davide Maiorca, Giorgio Giacinto, Claudia Eckert, and Fabio Roli. **Adversarial malware binaries: Evading deep learning for malware detection in executables**. In: *2018 26th European signal processing conference (EUSIPCO)*. IEEE. 2018, 533–537.

[Koz90]   John R Koza. **Non-linear genetic algorithms for solving problems**. *Google. Patents* (1990). US Patent 4,935,877.

[Krč+18]   Marek Krčál, Ondřej Švec, Martin Bálek, and Otakar Jašek. **Deep convolutional malware classifiers can learn from raw executables and labels only** (2018).

[Kre+18]   Felix Kreuk, Assi Barak, Shir Aviv-Reuven, Moran Baruch, Benny Pinkas, and
           Joseph Keshet. **Deceiving end-to-end deep learning malware detectors using
           adversarial examples.** *arXiv preprint* arXiv:1802.04528 (2018).

[KT+09]   Aleksander Kolcz, Choon-Hui Teo, et al. **Feature weighting for improved clas-
           sifier robustness.** In: Conference Organising Committee. 2009.

[Kum+20]   Ram Shankar Siva Kumar, Jonathon Penney, Bruce Schneier, and Kendra
           Albert. **Legal risks of adversarial machine learning research.** *arXiv
           preprint* arXiv:2006.16179 (2020).

[Lab+22]   Raphael Labaca-Castro, Luis Muñoz-González, Feargus Pendlebury, Gabi Dreo
           Rodosek, Fabio Pierazzi, and Lorenzo Cavallaro. **Realizable Universal Adver-
           sarial Perturbations for Malware.** *arXiv preprint* arXiv:2102.06747 (2022).

[Lab15]   Raphael Labaca-Castro. **Redefiniendo Límites de las Amenazas Digitales:
           Malware que Infecta Genomas Sintéticos Bacterianos.** MA thesis. Universi-
           dad de Buenos Aires, 2015.

[Lab22]   Raphael Labaca-Castro. **Framework for Adversarial Malware Evaluation
           Repository.** *GitHub* (2022). URL: https://github.com/zRapha/FAME (visited
           on 03/01/2022).

[Lar+21]   Héctor Laria, Yaxing Wang, Joost van de Weijer, and Bogdan Raducanu. **Hyper-
           GAN: Transferring Unconditional to Conditional GANs with HyperNet-
           works.** *arXiv preprint* arXiv:2112.02219 (2021).

[LBD19]   Raphael Labaca-Castro, Battista Biggio, and Gabi Dreo Rodosek. **Poster:
           Attacking malware classifiers by crafting gradient-attacks that preserve
           functionality.** In: *Proceedings of the 2019 ACM SIGSAC Conference on Com-
           puter and Communications Security.* 2019, 2565–2567.

[LeC+98]   Yann LeCun, Léon Bottou, Yoshua Bengio, and Patrick Haffner. **Gradientbased
           learning applied to document recognition.** *Proceedings of the IEEE* 86:11
           (1998), 2278–2324.

[LFR21]   Raphael Labaca-Castro, Sebastian Franz, and Gabi Dreo Rodosek. **AIMED-RL:
           Exploring Adversarial Malware Examples with Reinforcement Learning.**
           In: *Joint European Conference on Machine Learning and Knowledge Discovery
           in Databases.* Springer. 2021, 37–52.

[Lit+21]   Michael L. Littman, Ifeoma Ajunwa, Guy Berger, Craig Boutilier, Morgan Cur-
           rie, Finale Doshi-Velez, Gillian Hadfield, Michael C. Horowitz, Charles Isbell,
           Hiroaki Kitano, Karen Levy, Terah Lyons, Melanie Mitchell, Julie Shah, Steven
           Sloman, Shannon Vallor, and Toby Walsh. **Gathering Strength, Gathering
           Storms: The One Hundred Year Study on Artificial Intelligence (AI100)
           2021 Study Panel Report.** *Stanford University* 51 (2021). URL: http://ai100.
           stanford.edu/2021-report.

[LM05]   Daniel Lowd and Christopher Meek. **Adversarial learning.** In: *Proceedings of
           the eleventh ACM SIGKDD international conference on Knowledge discovery
           in data mining.* 2005, 641–647.

[LR22]   Raphael Labaca-Castro and Gabi Dreo Rodosek. **Framework for Adversarial
           Malware Evaluation.** In: *Journal under submission,* 2022.

[LS95]   James R Larus and Eric Schnarr. **EEL: Machine-independent executable edit-
           ing.** In: *Proceedings of the ACM SIGPLAN 1995 conference on Programming
           language design and implementation.* 1995, 291–300.

[LSR19a] Raphael Labaca-Castro, Corinna Schmitt, and Gabi Dreo Rodosek. **AIMED: Evolving Malware with Genetic Programming to Evade Detection**. In: *2019 18th IEEE International Conference On Trust, Security And Privacy In Computing And Communications/13th IEEE International Conference On Big Data Science And Engineering (TrustCom/BigDataSE)*. 2019, 240–247.

[LSR19b] Raphael Labaca-Castro, Corinna Schmitt, and Gabi Dreo Rodosek. **ARMED: How Automatic Malware Modifications Can Evade Static Detection?** In: *2019 5th International Conference on Information Management (ICIM)*. 2019, 20–27.

[Luc+21] Keane Lucas, Mahmood Sharif, Lujo Bauer, Michael K Reiter, and Saurabh Shintre. **Malware Makeover: Breaking ML-based Static Analysis by Modifying Executable Bytes**. In: *Proceedings of the 2021 ACM Asia Conference on Computer and Communications Security*. 2021, 744–758.

[Mad+17] Aleksander Madry, Aleksandar Makelov, Ludwig Schmidt, Dimitris Tsipras, and Adrian Vladu. **Towards deep learning models resistant to adversarial attacks**. *arXiv preprint* arXiv:1706.06083 (2017).

[Mat+18] Matteo Hessel, Joseph Modayil, Hado van Hasselt, Tom Schaul, Georg Ostrovski, Will Dabney, Dan Horgan, Bilal Piot, Mohammad Azar, and David Silver. **Rainbow: Combining Improvements in Deep Reinforcement Learning**. *Proceedings of the AAAI Conference on Artificial Intelligence* 32:1 (2018), 3215–3222. ISSN: 2374–3468.

[MLJ20] Markus F.X.J. Oberhumer, László Molnár, and John F. Reiser. **UPX: the Ultimate Packer for eXecutables—Homepage**. *GitHub* (2020). URL: https://upx.github.io/ (visited on 05/03/2021).

[Moo+17] Seyed-Mohsen Moosavi-Dezfooli, Alhussein Fawzi, Omar Fawzi, and Pascal Frossard. **Universal adversarial perturbations**. In: *Proceedings of the IEEE conference on computer vision and pattern recognition*. 2017, 1765–1773.

[Mop+18] Konda Reddy Mopuri, Utkarsh Ojha, Utsav Garg, and R. Venkatesh Babu. **NAG: Network for Adversary Generation**. In: *CVPR*. IEEE Computer Society, 2018, 742–751.

[Nee+19] Paarth Neekhara, Shehzeen Hussain, Prakhar Pandey, Shlomo Dubnov, Julian J. McAuley, and Farinaz Koushanfar. **Universal Adversarial Perturbations for Speech Recognition Systems**. In: *INTERSPEECH*. ISCA, 2019, 481–485.

[Nel+08] Blaine Nelson, Marco Barreno, Fuching Jack Chi, Anthony D Joseph, Benjamin IP Rubinstein, Udam Saini, Charles Sutton, J Doug Tygar, and Kai Xia. **Exploiting machine learning to subvert your spam filter**. *LEET* 8:1–9 (2008), 16–17.

[Ng+21] Edwin G Ng, Chung-Cheng Chiu, Yu Zhang, and William Chan. **Pushing the limits of non-autoregressive speech recognition**. *arXiv preprint* arXiv:2104.03416 (2021).

[Nor+09] Sadia Noreen, Shafaq Murtaza, M Zubair Shafiq, and Muddassar Farooq. **Evolvable Malware**. In: *11th Annual Conference on Genetic and Evolutionary Computation*. GECCO. New York, NY, USA: ACM, July 2009, 1569–1576.

[OCC13] Jonathan Oliver, Chun Cheng, and Yanggui Chen. **TLSH-locality sensitive hash**. In: *2013 Fourth Cybercrime and Trustworthy Computing Workshop*. IEEE. 2013, 7–13.

[Pie+20]   Fabio Pierazzi, Feargus Pendlebury, Jacopo Cortellazzi, and Lorenzo Cavallaro. **Intriguing properties of adversarial ML attacks in the problem space**. In: *2020 IEEE Symposium on Security and Privacy (S&P)*. IEEE. 2020,1332–1349.

[Pie94]   Matt Pietrek. **Peering Inside the PE: A Tour of the Win32 Portable Executable File Format**. *Microsoft Portal* (1994). URL: https://msdn.microsoft.com/en-us/library/ms809762.aspx (visited on 01/22/2022).

[Pos+20]   Richard Poschinger, Nils Rodday, Raphael Labaca-Castro, and Gabi Dreo Rodosek. **OpenMTD: A Framework for Efficient Network-Level MTD Evaluation**. In: *Proceedings of the 7th ACM Workshop on Moving Target Defense*. 2020, 31–41.

[Pra18]   Subhojeet Pramanik. **Malware detection using Convolutional Neural Networks Repository**. *GitHub* (2018). URL: https://github.com/subho406/Malware-detection-using-Convolutional-Neural-Networks (visited on 12/20/2021).

[PY20]   Daniel Park and Bülent Yener. **A survey on practical adversarial examples. for malware classifiers**. *arXivpreprint* arXiv:2011.05973 (2020).

[Qay+20]   Adnan Qayyum, Muhammad Usama, Junaid Qadir, and Ala Al-Fuqaha. **Securing connected & autonomous vehicles: Challenges posed by adversarial machine learning and the way forward**. *IEEE Communications Surveys & Tutorials* 22:2 (2020), 998–1026.

[Rad89]   Yisrael Radai. **The Israeli PC Virus**. *Computers & Security* 8:2 (1989), 111–113.

[Raf+17]   Edward Raff, Jon Barker, Jared Sylvester, Robert Brandon, Bryan Catanzaro, and Charles Nicholas. **Malware detection by eating a whole exe**. *arXiv preprint* arXiv:1710.09435 (2017).

[RCJ13]   Vaibhav Rastogi, Yan Chen, and Xuxian Jiang. **Droidchameleon: Evaluating. Android Anti-Malware Against Transformation Attacks**. In: *8th ACM SIGSAC Symposium on Information, Computer and Communications Security*. ASIACCS. New York, NY, USA: ACM, May 2013, 329–334.

[RG18]   Maria Rigaki and Sebastian Garcia. **Bringing a gan to a knife-fight: Adapting malware communication to avoid detection**. In: *2018 IEEE Security and Privacy Workshops (SPW)*. IEEE. 2018, 70–75.

[RLJ20]   Jonathan G Richens, Ciarán M Lee, and Saurabh Johri. **Improving the accuracy of medical diagnosis with causal machine learning**. *Nature communications* 11:1 (2020), 1–9.

[Ros+18]   Ishai Rosenberg, Asaf Shabtai, Lior Rokach, and Yuval Elovici. **Generic black. box end-to-end attack against state of the art API call based malware classifiers**. In: *International Symposium on Research in Attacks, Intrusions, and Defenses*. Springer. 2018.

[Roz05]   Konstantin Rozinov. **Reverse code engineering: An in-depth analysis of the bagle virus**. In: *Proceedings from the Sixth Annual IEEE SMC Information Assurance Workshop*. IEEE. 2005, 380–387.

[RS22]   Aqib Rashid and Jose Such. **StratDef: a strategic defense against adversarial attacks in malware detection**. *arXivpreprint* arXiv:2202.07568 (2022).

[SCJ19]   Octavian Suciu, Scott E Coull, and Jeffrey Johns. **Exploring adversarial examples in malware detection**. In: *2019 IEEE Security and Privacy Workshops (SPW)*. IEEE. 2019, 8–14.

[Sev+21]   Giorgio Severi, Jim Meyer, Scott Coull, and Alina Oprea. **Explanation-Guided Backdoor Poisoning Attacks Against Malware Classifiers**. In: *30th USENIXSecurity Symposium (USENIXSecurity 21)*. 2021, 1487–1504.

[SH82]   John F Shoch and Jon A Hupp. **The "worm" programs-early experience with a distributed computation**. *Communications of the ACM* 25:3 (1982), 172–180.

[ŠL13]   Nedim Šrndic and Pavel Laskov. **Detection of malicious pdf files based on hierarchical document structure**. In: *Proceedings of the 20th Annual Network & Distributed System Security Symposium*. Citeseer. 2013, 1–16.

[Sri93]   Amitabh Srivastava. **A practical system for intermodule code optimization at link-time**. *Journal of programming Languages* 1:1 (1993), 1–18.

[SS12]   Charles Smutz and Angelos Stavrou. **Malicious PDF detection using metadata and structural features**. In: *Proceedings of the 28th annual computer security applications conference*. 2012, 239–248.

[Sze+13]   Christian Szegedy, Wojciech Zaremba, Ilya Sutskever, Joan Bruna, Dumitru Erhan, Ian Goodfellow, and Rob Fergus. **Intriguing properties of neural networks**. *arXiv* (2013). URL: http://arxiv.org/pdf/1312.6199v4.

[Tho18]   Romain Thomas. **LIEF—Library to Instrument Executable Formats Repository**. *GitHub* (2018). URL: https://github.com/lief-project/LIEF (visited on 08/31/2018).

[Tob+16]   Shun Tobiyama, Yukiko Yamaguchi, Hajime Shimada, Tomonori Ikuse, and Takeshi Yagi. **Malware detection with deep neural network using process behavior**. In: *2016 IEEE 40th annual computer software and applications conference (COMPSAC)*. Vol. 2. IEEE. 2016, 577–582.

[Tra+19]   Florian Tramèr, Pascal Dupré, Gili Rusak, Giancarlo Pellegrino, and Dan Boneh. **Adversarial: Perceptual Ad Blocking Meets Adversarial Machine Learning**. In: *Conference on Computer and Communications Security*. 2019, 2005–2021.

[Ues+18]   Jonathan Uesato, Brendan O'donoghue, Pushmeet Kohli, and Aaron Oord. **Adversarial risk and the dangers of evaluating against weak attacks**. In: *International Conference on Machine Learning*. PMLR. 2018, 5025–5034.

[UZ15]   Jason Upchurch and Xiaobo Zhou. **Variant: a malware similarity testing framework**. In: *201510th International Conference on Malicious and Unwanted Software (MALWARE)*. IEEE. 2015, 31–39.

[VGS16]   Hado Van Hasselt, Arthur Guez, and David Silver. **Deep reinforcement learning with double q-learning**. In: *Proceedings of the AAAI conference on artificial intelligence*. Vol. 30. 1. 2016.

[Vir21a]   VirusShare. **VirusShare**. *VirusShare* (2021). URL: https://virusshare.com (visited on 03/12/2021).

[Vir21b]   VirusTotal. **VirusTotal.com: Analyze suspicious files and URLs to detect types of malware, automatically share them with the security community**. *VirusTotal* (2021). URL: https://www.virustotal.com/ (visited on 04/27/2021).

[Vir21c]   VirusTotal. **VirusTotal.com: AV product on VirusTotal detects a file and its equivalent commercial version does not**. *VirusTotal* (2021). URL: https://support.virustotal.com/hc/en-us/articles/115002122285-AV-product-on-VirusTotal-detects-a-file-and-its-equivalent-commercial-version-does-not (visited on 05/05/2021).

[Vir22]   VirusTotal. **VirusTotal.com: Binary Analysis**. *VirusTotal* (2022). URL: https:// www.virustotal.com/gui/search/malware/ (visited on 02/02/2022).

[Vis+21]   Corrado Aaron Visaggio, Fiammetta Marulli, Sonia Laudanna, Benedetta La Zazzera, and Antonio Pirozzi. **A Comparative Study of Adversarial Attacks to Malware Detectors Based on Deep Learning**. In: *Malware Analysis Using Artificial Intelligence and Deep Learning*. Springer, 2021, 477511.

[WD92]   Christopher J. C. H. Watkins and Peter Dayan. **Q-learning**. *Machine learning* 8 (1992) (1992), 279–292. https://doi.org/10.1007/BF00992698.

[Wil+16]   Mark D Wilkinson, Michel Dumontier, Ijsbrand Jan Aalbersberg, Gabrielle Appleton, Myles Axton, Arie Baak, Niklas Blomberg, Jan-Willem Boiten, Luiz Bonino da Silva Santos, Philip E Bourne, et al. **The FAIR Guiding Principles for scientific data management and stewardship**. *Scientific data* 3:1 (2016), 1–9.

[Xia+18]   Chaowei Xiao, Ruizhi Deng, Bo Li, Fisher Yu, Mingyan Liu, and Dawn Song. **Characterizing adversarial examples based on spatial consistency information for semantic segmentation**. In: *Proceedings of the European Conference on Computer Vision (ECCV)*. 2018, 217–234.

[XQE16]   Weilin Xu, Yanjun Qi, and David Evans. **Automatically evading classifiers**. In: *Proceedings of the 2016 network and distributed systems symposium*. Vol. 10. 2016.

[Yan+17]   Wei Yang, Deguang Kong, Tao Xie, and Carl A Gunter. **Malware detection in adversarial settings: Exploiting feature evolutions and confusions in android apps**. In: *Proceedings of the 33rd Annual Computer Security Applications Conference*. 2017, 288–302.

[YPT22]   Javier Yuste, Eduardo G Pardo, and Juan Tapiador. **Optimization of code caves in malware binaries to evade Machine Learning detectors**. *Computers & Security* (2022), 102643.

[Yua+20]   Junkun Yuan, Shaofang Zhou, Lanfen Lin, Feng Wang, and Jia Cui. **Blackbox adversarial attacks against deep learning based malware binaries detection with GAN**. In: *ECAI 2020*. IOS Press, 2020, 2536–2542.

[ZD16]   Ziyun Zhu and Tudor Dumitraş. **Featuresmith: Automatically engineering features for malware detection by mining the security literature**. In: *Proceedings of the 2016 ACM SIGSAC Conference on Computer and Communications Security*. 2016, 767–778.

[Zha+20]   Yu Zhang, James Qin, Daniel S Park, Wei Han, Chung-Cheng Chiu, Ruoming Pang, Quoc V Le, and Yonghui Wu. **Pushing the limits of semi-supervised learning for automatic speech recognition**. *arXiv preprint* arXiv:2010.10504 (2020).

[ZLC20]   Lan Zhang, Peng Liu, and Yoon-Ho Choi. **Semantic-preserving Reinforcement Learning Attack Against Graph Neural Networks for Malware Detection**. *arXiv preprint* arXiv:2009.05602 (2020).

[ZLL12]   Min Zheng, Patrick PC Lee, and John CS Lui. **ADAM: an automatic and extensible platform to stress test android anti-virus systems**. In: *International conference on detection of intrusions and malware, and vulnerability assessment*. Springer. 2012, 82–101.

Printed in the United States
by Baker & Taylor Publisher Services